Synthesis Lectures on Biomedical Engineering

Series Editor

John Enderle, Storr, USA

This series consists of concise books on advanced and state-of-the-art topics that span the field of biomedical engineering. Each Lecture covers the fundamental principles in a unified manner, develops underlying concepts needed for sequential material, and progresses to more advanced topics and design. The authors selected to write the Lectures are leading experts on the subject who have extensive background in theory, application, and design. The series is designed to meet the demands of the 21st century technology and the rapid advancements in the all-encompassing field of biomedical engineering.

Khalid Sayood · Hasan H. Otu

Bioinformatics

A One Semester Course

 Springer

Khalid Sayood
Department of Electrical and Computer
Engineering
University of Nebraska–Lincoln
Lincoln, NE, USA

Hasan H. Otu
Department of Electrical and Computer
Engineering
University of Nebraska–Lincoln
Lincoln, NE, USA

ISSN 1930-0328   ISSN 1930-0336 (electronic)
Synthesis Lectures on Biomedical Engineering
ISBN 978-3-031-20016-8   ISBN 978-3-031-20017-5 (eBook)
https://doi.org/10.1007/978-3-031-20017-5

This Springer imprint is published by the registered company Springer Nature Switzerland AG
The registered company address is: Gewerbestrasse 11, 6330 Cham, Switzerland

Contents

Introduction

1

In a nutshell ... We look at some of the different definitions of bioinformatics and try to identify some of the major areas that make up the field of bioinformatics.

1.1   What Is Bioinformatics?

The definition of bioinformatics is fluid and changes depending on who we ask. To many people, it is any use of computers to handle biological information. To others, it is the development of databases and query tools used for the exploration of genomic sequences. Here is a formal definition which reflects these views [1]:

> Bioinformatics is conceptualizing biology in terms of molecules (in the sense of physical chemistry) and applying informatics techniques (derived from disciplines such as applied mathematics, computer science, system analysis, and statistics) to understand and organize the information associated with those molecules, on a large scale.

Another somewhat restrictive definition was provided by the NIH Biomedical Information Science and Technology Initiative (BISTI) Consortium [2]:

> Research, development, or application of computational tools and approaches for expanding the use of biological, medical, behavioral or health data, including those to acquire, store, organize, archive, analyze, or visualize such data.

BISTI also defined *computational biology* as:

> The development and application of data-analytical and theoretical methods, mathematical modeling, and computational simulation techniques to the study of biological, behavioral, and social systems.

© The Author(s), under exclusive license to Springer Nature Switzerland AG 2022   1
K. Sayood and H. H. Otu, *Bioinformatics*, Synthesis Lectures on Biomedical Engineering,
https://doi.org/10.1007/978-3-031-20017-5_1

We define bioinformatics as the study of the information contained in biological molecules, such as DNA and proteins, in their interactions, and in the systems built by these molecules. As form fits function, this definition of bioinformatics means that the study of bioinformatics needs to include an understanding of the biological processes which influence the organization of information in biological systems. Our definition will guide our study but we will find that the subjects we study also form the core of the subjects influenced by other definitions.

1.2   Foundations of Bioinformatics

The reason for the diversity of definitions is the relative youth of the field of bioinformatics. Much of what we study in bioinformatics is only possible because of the synergistic interaction between technological advancements and the (sometimes resultant) advances in our understanding of biological processes. The exponential growth of computational power and storage when married to the breakthroughs in our understanding of how information is organized and communicated in living organisms has resulted in the generation of large and complex datasets that require specific management and analysis. This has shaped the birth and growth of bioinformatics.

As shown in Fig. 1.1, three domains drive the development of bioinformatics. The subject matter means that one of the domains is the field of life sciences. The dependence on technol-

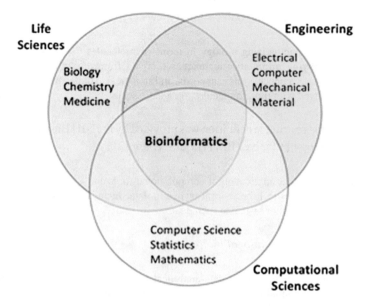

Fig. 1.1 Disciplines utilized by bioinformatics

ogy means that another domain that influences our study consists of engineering fields. And because we have to extract information from large volumes of data, this necessarily brings in the domain of computational sciences. In particular, the life sciences discipline include biology, chemistry, and medicine; the engineering disciplines include electrical, mechanical, materials, and computer engineering; and the computational sciences disciplines include computer science, mathematics, and statistics.

For example, using biochemical techniques from the life sciences that involve polymerase chain reaction (PCR), restriction enzymes, or chemical digestion, macromolecules such as DNA can be selected and amplified in a way that can be detected and quantified by specialized machines such as sequencers, microarrays, or mass spectrometers. These machines are developed based on advanced engineering principles and generate high-throughput, reliable data, such as genome sequences, gene expression levels, or protein abundance. In turn, this large data is annotated, stored, and analyzed with advanced ontologies, databases, and algorithms using techniques rooted in the computational sciences.

In order to have any kind of handle on this vast and ever expanding field, we need to partition this enormous field into smaller (though still large) specific areas. It is difficult to describe the specific areas bioinformatics deals with without first describing the biological concepts that are involved in those areas. Nevertheless, here is an attempt to list the main areas studied in bioinformatics with brief descriptions, followed by slightly more detailed descriptions of particular areas.

(Comparative) Genome analysis: The genome of an organism represents the entirety of the DNA found in its cell. This includes the chromosomes as well as extra-chromosomal sequences such as plasmids in bacteria, choloroplast DNA in plants, and mitochondrial DNA in most eukaryotes. This area includes sequence assembly, which is the reconstruction of the entire genome from its parts; genome annotation, which involves the identification and characterization of genomic features, such as repeat regions, insertions/deletions, genes, and promoters; epigenetics, which deals with heritable, functionally important changes in the genome that do not change the sequence of the genome, e.g., DNA methylation; and comparison of genome characterization and structure between organisms to understand evolution, population dynamics, and genome functionality.

Omics analysis: The word/suffix "omics," is a neologism that is used to represent the study of the ensemble of a specific type of molecule in biology. Genomics studies the genome and changes in it, e.g., mutations, structural variations; transcriptomics involves gene expression levels, i.e., levels or numbers of RNA transcripts; proteomics assesses protein abundance. Such definitions can be extended to other "omes" such as the metabolome—all metabolites in a biological sample, or lipidome—all lipids in a cell—which is in turn part of the metabolome. The goal in omics is to derive associations and causal relations between the omic or ensemble levels and phenotype. For example, a genomic study in breast cancer tries to identify the mutations specific to the genomes of breast cancer patients whereas a proteomic study aims at assessing the difference in protein abundance in samples from breast cancer patients when compared with samples from people without breast cancer. The next step in this area is "multiomic analysis," which aims to provide a holistic approach to the analysis of different omic types.

Meta-omics: This area involves the cases when the omics data come from a collection of organisms. Meta-omics is typically applied in microbiology for the analysis of communities of bacteria, archaea, yeast, and viruses. The most common approach in this area is metagenomics, which aims to analyze the microbial species in a community by sequencing and assembling the DNA found in the community.

Sequence analysis: The goal in this area is to use DNA/RNA/protein sequences from different sources to infer similarity, e.g., using sequence alignment, identify functional subunits in these molecules, e.g., motifs in DNA or domains in proteins, find genes or proteins through computational analysis of new and known sequence data, perform phylogenetic analysis reconstructing evolutionary histories and mechanisms based on molecular sequences, predict the function of new sequences, and search and store large sequence collections.

Structural studies: Although our initial approach to understanding the biological molecules DNA/RNA/protein is through their sequence composition (called the primary sequence/structure), these molecules exist in three-dimensional (3D) space and the structure of a molecule in 3D space, (tertiary structure) critically affects its function. Furthermore, there is a quaternary aspect to the structural studies that involve protein-protein or protein-DNA interactions for known 3D structures. Structural bioinformatics deals with predicting the 3D structure of molecules using their primary structure (sequence information), which has implications in protein docking, drug design, functional prediction, and protein engineering.

Systems/Network Biology: The interaction of biological entities at the molecular, cellular, or organismal levels, can be analyzed through network science. This area involves the study of biological pathways, functional groups of genes, and the interaction and regulation networks that explain the mechanisms underlying observed phenotypic states and population dynamics.

Signal and Image Analysis: This area deals with signal and image acquisition, processing, pattern recognition, and annotation of data resulting from approaches that involve technologies such as microscopy, magnetic resonance imaging (MRI), electroencephalogram (EEG), and electrocardiogram (ECG or EKG).

Information/Knowledge Processing: This area represents a wide array of approaches that are not rooted specifically in bioinformatics but are general informatic approaches applied to data studied in bioinformatics. These approaches include but are not limited to text mining, ontology development, database schemas, data visualization techniques, hardware/software architecture, and cybersecurity methodologies that are geared towards data generated in life sciences.

Let's look at a few of these areas in slightly more detail.

Comparative Genomics

With the increasing availability of genomic sequences, it has become possible to draw inferences about how species evolved. Looking at how genes for the same function differ in different organisms can tell us about the relatedness of organisms. To perform comparisons, we need to develop methods of sequence comparisons and develop measures of differences between sequences. These tasks are not trivial. Firstly, there is the problem of *gene finding*. A great deal of effort has been spent in developing algorithms for gene finding. Once you have a gene for a particular species you need to find a homologous sequence in other species. Or, you might determine if a sequence is a gene by looking to see if the same sequence has been identified as a gene in other studies. In either case, this means searching for a match through vast databases in a reasonable amount of time. Not only that, by the very nature of the search the match is not exact. Thus the search results need to include sequences that are "somewhat" similar. Once the search has provided a set of results, these results have to be compared in a unified framework. To perform comparisons, there needs to exist a measure of similarity or difference. This measure of similarity or difference should not only be computable, it should also reflect biological distance. All this falls under the title of *sequence comparison*. We will be spending some time looking at various approaches to comparing sequences.

We mentioned that the difference between sequences should reflect biological distance. What does biological distance mean? It could mean a measure of the evolutionary relationship. How long has it been since the organisms in question descended from a common ancestor? If we are talking about multiple organisms, which diverged first? If we have genes from two different organisms which we determine to be very similar, does that mean they are evolutionarily related or did the genes become similar due to convergent evolution? In other words are the genes homologous. If they are not, we should not be asking questions about evolutionary distance. A distance or similarity measure is useful only if the results are biologically meaningful.

Genes interact with each other to form regulatory networks. An important component of these networks are regulatory sites that enhance or depress gene expression. Differences– and similarities—between regulatory regions of homologous genes can give us clues to the evolutionary relationships between organisms. An important bioinformatics task is the development of computational tools for the discovery of regulatory regions.

Notice that throughout this description we are talking about computational techniques that are precise, and definitive, to be used to obtain results which are not precisely defined. Herein lies one of the problems of bioinformatics. The people used to dealing with computational techniques and algorithms generally do not have enough knowledge to feel very comfortable with the meaning of the results of their computations. People who are familiar with the biological meaning often do not have a feel for the capabilities and limitations of the computational approaches. This tension is at the heart of many interactions between computational and biological scientists and needs to be recognized and acknowledged.

Functional Genomics

Functional genomics is the study of the function and interaction of genes and the products of the genes. These products include proteins as well as RNA sequences that do not get translated into proteins. While it is relatively straightforward to obtain the primary protein sequence from the gene, it is much more difficult to infer the function. In part, this is because the function often depends on the three dimensional tertiary and quaternary structure of the protein rather than the linear primary structure. Even when the three-dimensional structure of the protein can be predicted with some degree of accuracy, it is still a non-trivial matter to deduce the actual function of the protein. High throughput techniques like the use of microarrays, or RNASeq techniques involving the sequencing of transcriptomes, produce vast quantities of data that can be used to relate the expression of various genes and hence get an understanding of the interactions between genes and gene products. The amount of data produced by high throughput approaches is at the same time too much and too little. It is too much in the sense that it is difficult to make sense of such a large quantity of data. It is too little in that the data generated is a very limited snapshot of a hugely complex process. We will look at various approaches to extracting the functions of genes and the functional relationships between genes from high throughput data. The relationships between genes can also be studied by identifying regulatory mechanisms. These regulatory mechanisms are mediated through the use of transcription factors and regulatory sites which are embedded in the DNA sequence. Explicating these mechanisms is one of the tasks of bioinformatics.

Proteomics

Beadle and Tatum in 1941 [3] came up with the one gene-one protein hypothesis for which (in part) they received the Nobel prize. From that, it would seem that by knowing the genes we could obtain all the information we need about the protein composition within an organism. However, it turns out that reality is somewhat more complicated. Apart from the fact that the one gene-one protein hypothesis is not entirely true, especially in eukaryotes, the fact that an organism has a gene for a particular protein does not mean that it will express that protein. Different organisms carrying the same gene will express the protein at different levels. In fact, the same organism will express the protein at different levels, or not at all, at different times and under different circumstances. Finally, many proteins will get modified through a process called post-translational modification before they are used. Thus the importance of studying proteins along with the genome. The word proteome is a combination of the first part of the word "protein" and the second part of the word "genome." It is the set of proteins produced by the organism during its life.

Structural Genomics

Most biological molecules do not exist as linear molecules but have a three-dimensional structure. It is this three-dimensional structure that allows them to perform their required functions. The study of the three-dimensional structure of proteins falls under the heading of structural genomics. The primary method of obtaining the three dimensional structure of proteins is through experimental techniques such as x-ray crystallography. However, experimental techniques are expensive, time consuming, and not always feasible. Therefore, there is increasing interest in the use of computational techniques to obtain the three-dimensional structure of proteins.

Metagenomics

The evolutionary scientist and prolific author Steven Jay Gould [4] pointed out that the standard illustration in popular articles about evolution which show various ages, such as the age of the dinosaur, etc. are very misleading. There has, in fact, only been one age, one that continues to this day, and that is the age of the bacteria. The world was, is, and will be colonized by bacteria. Even when it comes to human beings, the number of bacteria carried by a human being is estimated to be more than the number of cells in the human body. To the dominant species in the world, human beings may be just another form of transportation and nutrition. The environment around us including the soil, the air, and the water contains trillions of bacteria many involved in activities beneficial to humans, such as recycling and absorbing the carbon dioxide in the environment. Despite the importance and ubiquity of bacteria, it is estimated that we have knowledge of less than 1% of bacterial species. One reason for this is that to study a particular species of bacteria, it is often required that we grow them in culture. Most bacteria live in collaborative communities making it difficult, if not impossible, to grow them in isolation. Metagenomics is the study of the genomes and the genetic material of bacteria living in colonies. Given that we cannot isolate these bacteria from each other, there is a need for new approaches to studying these bacteria and their impact on humans. Many of these new approaches are bioinformatic approaches.

1.3   Organization of This Book

In this book, we will mostly restrict ourselves to the study of a few of the tools that are widely used in the areas of comparative genomics, functional genomics, and metagenomics. As one of our motivating principles is that form fits function, we also present some of the basic ideas in molecular biology. Hopefully, these ideas will give us a context for the bioinformatics tasks that await us. We will start with a look at the function and hopefully use the clues we get to look at the form. We will begin the course by looking at the chemical structure of biological molecules. The idea is more to build up the vocabulary we will need to talk about these molecules than to become experts in the chemistry of these molecules. We will then look at the biological processes which leave their imprint on the genome. This includes replication, transcription, and post-transcriptional and post-translational processes such as splicing, mRNA degradation, and post-translational modifications. Each of these processes imposes different kinds of organizational and structural constraints on the DNA. In the process, we will also look at various bioinformatics tools available for identifying some of the components of the genome. We will then study the different ways in which we get genomic variations including mutation and recombination, and then look at methods of sequence alignment which depend on mathematical models of evolution. We will look at some of the most widely used methods of both local and global alignments. In particular, we will study the BLAST algorithm which is the most widely used tool in bioinformatics. Coming full circle we will use the sequence alignment methods to begin our study of molecular phylogeny.

We have said earlier that one of the main goals of Bioinformatics is to understand how information is organized in the DNA. There are at least millions of different kinds of organisms each of which has come up with its own answer to how information can be organized. This organization itself results in the generation of a signature of the species. We will study different ways of obtaining signatures and then connect that to molecular phylogeny.

1.4   Exercises

1. One of the main repositories for Bioinformatics is the National Center for Biotechnology Information (NCBI). Explore the NCBI website. What are the institutes and organizations that maintain NCBI?

2. List three databases/resources housed by NCBI. Describe the use and provide relevant statistics or history, where applicable, for each database/resource.

3. One of the databases hosted at NCBI is the dbSNP database. Explain what a SNP is. In a typical human genome, how many SNPs would one expect to observe?

4. Ensembl is a popular genome browser that offers many tools and data sets for Bioinformatics analysis. Visit the Ensmbl website and locate the main page for the Human genome. Go to the page that provides more information and statistics about the Human genome. How many coding genes are there on the Human genome? What is the Human genome's length, i.e., the number of base pairs of the Human genome?

5. Another popular genome browser is that of University of California, Santa Cruz's (UCSC). Visit the web page for the UCSC Genome Browser. Locate the UCSC Species Tree, which can be found on the main page for the Genome Browser menu item. List two viruses listed on this tree. What is the closest organism to Mouse based on this species tree?

6. On the UCSC Genome Browser website, search the term VEGFA in the Human genome. VEGFA stands for the "vascular endothelial growth factor A" gene. In the results page for your search, locate the item that says VEGFA and click on the link. Explore the UCSC genome browser view. Can you identify the chromosome VEGFA gene resides on? What is the location of the gene on this chromosome?

References

1. N.M. Luscombe, D. Greenbaum, and M. Gerstein. What is bioinformatics? a proposed definition and overview of the field. *Methods of Information in Medicine*, 40:346 – 358, 2001.
2. J. R. Smith. Bioinformatics and the eye. *J Ocul Biol Dis Infor*, 2(4):161–163, 2010.
3. George W Beadle and Edward L Tatum. Genetic control of biochemical reactions in neurospora. *Proceedings of the National Academy of Sciences*, 27(11):499–506, 1941.
4. Stephen Jay Gould. *Ontogeny and phylogeny.* Harvard University Press, 1985.

Molecular Biology Primer

2

In a nutshell .. We look at some of the basic concepts of molecular biology that will be of interest to us. Beginning with a definition of bonds we work our way to the introduction of the DNA sequence and some of its oddities.

In the beginning there was carbon—at least as far as life on earth is concerned. We are carbon-based life forms. Our bodies are made up of carbon compounds. Proteins, fats, sugars, etc. are all carbon compounds. However, it is not just the building blocks of the body that are based on carbon. The DNA molecules that contain the information from which our bodies are constructed are also carbon compounds. Why carbon? Why not something else? The answer lies in the requirements for molecules to perform the incredibly complex task of storing and preserving information over long periods. To store complex information, we need complex molecules. To store a lot of information, we need big molecules. To preserve this information over extended periods of time, we need relatively stable molecules. To build complex molecules, we need to be able to make multiple bonds; for these molecules to be large and stable, we need these bonds to be strong. The carbon atom satisfies all these requirements for construction of such molecules and has been the backbone of life on earth.

Before we look to see how well Carbon satisfies these requirements let's briefly review the various types of bonds.

2.1 Bonds

There are several ways elements bond to each other. Most of these result from the desire of elements to have a filled up set of orbitals—two electrons per orbital. The types of bonds of most interest to us are

- Ionic bonds.
- Covalent bonds.
- Polar bonds.

2.1.1   Ionic Bonds

Ionic bonds are formed when one atom gives up electrons resulting in a positively charged ion and another atom takes up electrons resulting in a negatively charged ion. The electrostatic attraction between these oppositely charged ions is the bond that holds these atoms together. The ionic bond is a very strong bond in vacuum. However, in aqueous solutions, it can fall apart. An example is Sodium Chloride. Sodium has eleven electrons, two in the first shell, eight in the second shell, and one in the third shell. Giving up this one electron leaves it with eight electrons in its outermost shell and a positive charge. Chlorine has seventeen electrons, two in the first shell, eight in the second shell, and seven in the outermost shell. By picking up the electron donated by Sodium it also has eight electrons in its outermost shell and a negative charge. The negatively charged Chlorine and the positively charged Sodium form an ionic bond to make Sodium Chloride. Outside of water this is a strong bond and Sodium and Chlorine stick together, however, put Sodium Chloride in water and we have a different story—about which more later.

2.1.2   Covalent Bonds

Covalent bonds are formed when atoms share electrons to achieve a more stable configuration. Electrons surround atoms in shells with each shell able to accommodate a particular number of electrons. The electrons exist in orbitals where an orbital is essentially the probability density function for the location of the electron. The first shell can have a maximum of two electrons, the second shell a maximum of eight electrons, and so on. The greatest stability is reached when the outer shell, or valence shell, has the maximum number of electrons it can accommodate. Hydrogen has a single shell with a single electron in it. As one might expect, the orbital is spherical. Two hydrogen atoms will therefore share their electrons. This allows both hydrogen atoms to (kind of) have the maximum number of electrons in their valence shell. The electrons in the valence shell are attracted to both nuclei thus keeping the two constituent atoms together. The orbital is in the shape of an oblong sausage. The bond formed by sharing electrons is called a covalent bond. Oxygen has eight electrons, two in the first shell and six in the outer shell. By sharing two electrons with two Hydrogen atoms it completes its outer shell while also completing the outer shells of the two Hydrogen atoms. This is how a water molecule is formed.

2.1.3 Polar Bonds

In a covalent partnership, all partners are not necessarily equal. In a water molecule, the
Oxygen atom has a larger nucleus with more positive charge so it exerts a greater pull on
the shared electrons than do the Hydrogen atoms. This makes the Oxygen end of the water
molecule more negative and the Hydrogen end more positive. This polar nature of water
allows it to interact with other charged particles. The negative Oxygen end can attract a
positively charged particle forming a bond with it, while the positive Hydrogen atoms can
form a bond with negatively charged particles. This is what happens when we put salt in
water. The positively charged Hydrogen ends pull on the negatively charged Chlorine while
the negatively charged Oxygen end pulls on the positively charged Sodium overcoming the
ionic bond.

Given the relatively small attraction of the hydrogen nucleus, covalent bonds involving
Hydrogen generally result in polar molecules. Bonds between these polar Hydrogens and
positively charged members of other covalent bonds are called Hydrogen bonds.

2.2 Carbon

So, now that we know what the different kinds of bonds are available, lets get back to
the question of why Carbon? First, while both ionic bonds and covalent bonds can be very
strong in vacuum, in aqueous solutions covalent bonds are much stronger. In fact, in aqueous
solutions covalent bonds are the strongest bonds possible. Carbon, with four electrons in
its outer shell forms four covalent bonds; enough to form some rather complex molecules.
That does not completely answer the question as there are other elements that can form
as many covalent bonds. Silicon, for example, can also form four covalent bonds. So what
distinguishes Carbon from Silicon? In Table 2.1 we have the bond energies for several
possible candidates

Notice how much stronger the Carbon–Carbon bond is than Silicon–Silicon. Furthermore,
the bond between two Carbons is stronger than the bond between Carbon and Oxygen. In
the case of Silicon, the Silicon–Silicon bond is weaker than the Silicon–Oxygen bond.
Therefore, if an Oxygen happens to come along it can easily disrupt any Silicon chain that

Table 2.1 Bond strengths in kJ/mol [1]

	X–X (Single bond)	X == X (double bond)	X – O
Carbon	347–356	611	336
Silicon	230		368
Oxygen	146	498	
Nitrogen	163	418	

might be attempting to grow. This would not be true for Carbon. Thus Carbon is ideally suited for making stable, long and complex chains. Which is what we need if we are to encode large amounts of complex information.

The study of carbon compounds, especially compounds containing Carbon Hydrogen bonds is so important that it forms a branch of Chemistry called Organic Chemistry. By adding different functional groups to Carbon–Hydrogen chains we get different classes of organic compounds.

2.3 Sugar

A *Sugar* is a carbohydrate with the generic formula $C_m(H_2O)_n$ and containing either an aldehyde group (O=CH) or a ketone group (C(C=O)C). For example, Glucose is a six Carbon sugar with an aldehyde group, and Fructose is a six Carbon sugar with a ketone group. Of particular interest to us is a five Carbon sugar with an aldehyde group called *ribose*.

The sugar ribose is a five Carbon sugar with an aldehyde functional group. The sugar can exist in both linear and cyclic forms. The Carbons in the sugar are numbered starting with the Carbon belonging to the aldehyde group as shown in Fig. 2.1. The ribose sugar provides the **R** of RNA, the ribonucleic acid.

If we remove one of the oxygens from ribose as shown in Fig. 2.2 we get the sugar deoxyribose (ribose without an oxygen). This deoxyribose provides the **D** of DNA, the deoxyribonucleic acid. In order to get the other two letters of RNA and DNA we have to add some stuff to this sugar. One of the structures we add to the sugar is a *nucleobase*, more commonly referred to as a *base*.

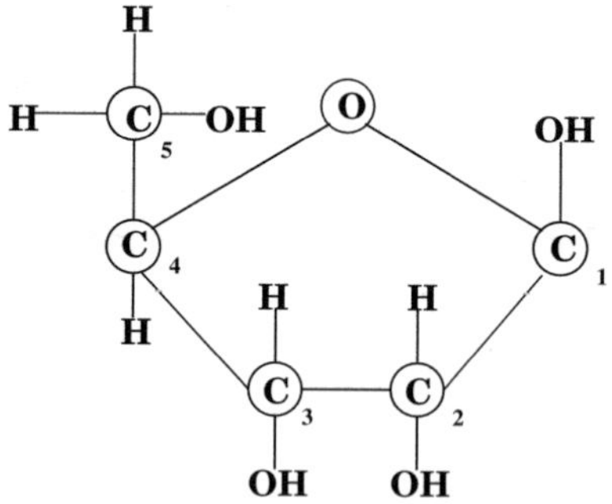

Fig. 2.1 The ribose sugar with the Carbon atoms numbered from 1 through 5

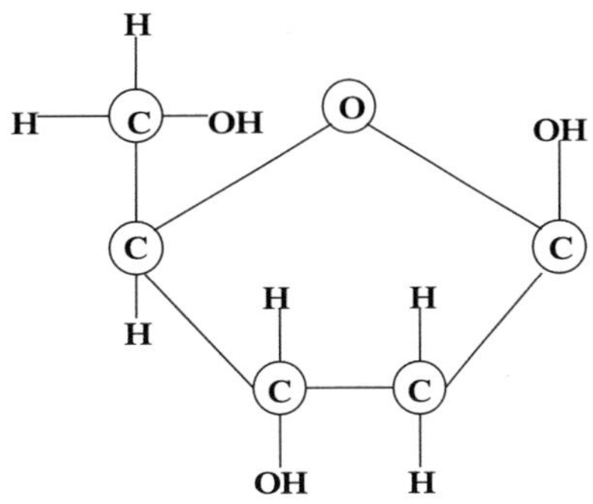

Fig. 2.2 The deoxyribose
sugar

2.3.1 Base

There are five different kinds of bases Adenine, Guanine, Cytosine, Thymine, and Uracil. Of these, the first four are used for building DNA. In RNA Uracil replaces Thymine. For the moment, let's concentrate on the first four. The structures of the four are shown in Fig. 2.3. We can see that Adenine and Guanine have two ring structures while Cytosine and Thymine have single ring structures. The single ring structure is representative of a group of organic compounds known as *pyrimidines*. Pyrimidines are six-atom rings with Nitrogen at positions 1 and 3. A well-known member of the Pyrimidine family is Benzene, a constituent of crude oil. Adenine and Guanine belong to a family called Purine. A Purine is made up of a Pyrimidine ring attached to a pentagonal ring called an Imidazole ring. These bases are what keep the two strands of the DNA together. And they do that using a weak polar bond. If we pair up the bases as shown in Fig. 2.4, we see that there are only two pairings that work. Adenine paired with Thymine and Cytosine paired with Guanine. Only in these pairings do the hydrogen bond donors and acceptors match up. These pairings are called the Chargaff rules, named after Erwin Chargaff who first inferred this relationship.

2.3.2 Nucleotide

If we take the ribose sugar and add one of a group of nucleobases to the 1 Carbon we get a *nucleoside*. If we then attach a polyphosphate group to the 5 Carbon we get a nucleotide. If instead of a ribose sugar we use a deoxyribose sugar to which we attach a nucleobase and a polyphosphate group we get a deoxynucleotide. We will refer to both as nucleotides as long as which one we mean is evident from the context. The nucleobases, or bases, are important in defining the structure and functioning of DNA. The joining of these nucleotides into a chain requires energy. This energy is provided by the polyphosphate group (Fig. 2.5).

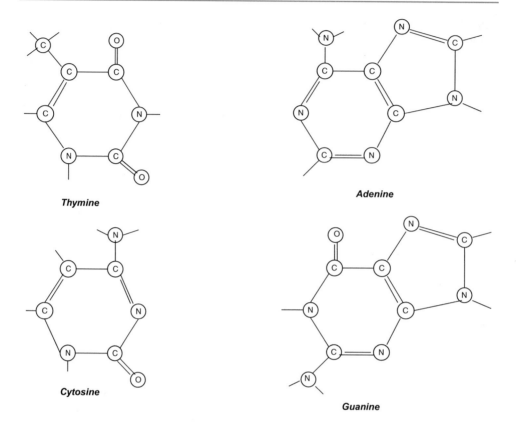

Fig. 2.3 Nucleobases

2.3.3 Phosphate Group

Biology is dynamic. Biological molecules like DNA and protein participate in numerous interactions which require the expenditure of energy. This includes the construction of the DNA and RNA molecules. In biological systems one of the major stores of energy are polyphosphates. Polyphosphates are polymers constructed by connecting PO_4 units through shared Oxygen atoms. These covalent bonds when broken provide energy required by the various biological molecules for their interactions. As shown in Fig. 2.6, the hydroxyl group (OH) at the 3rd Carbon (referred to as the 3' location) of the nucleotide at the top binds to the alpha-phosphate group at the 5th Carbon (referred to as the 5' location) of the nucleotide at the bottom to form a phosphodiester bond. The energy required to complete this reaction is provided by the breaking of the polyphosphate bond. Notice that at this point the dinucleotide has two distinct ends. A 5' end with the phosphate group attached and a 3' end. If we wish to attach another nucleotide to this chain, we will need to make use of the energy from the

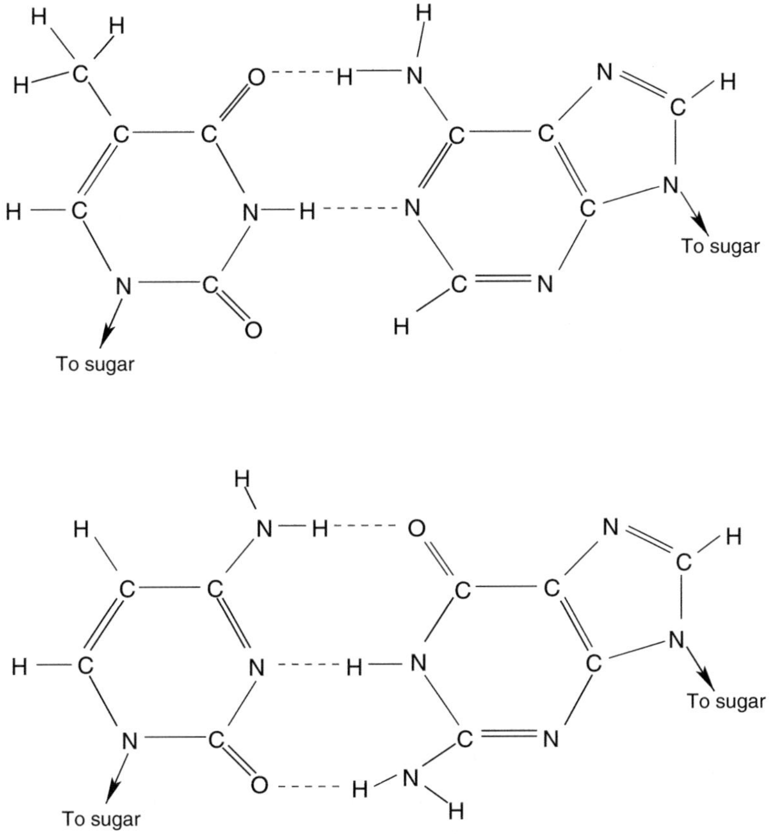

Fig. 2.4 Base pairing according to the Chargaff rules

Fig. 2.5 Nucleotide

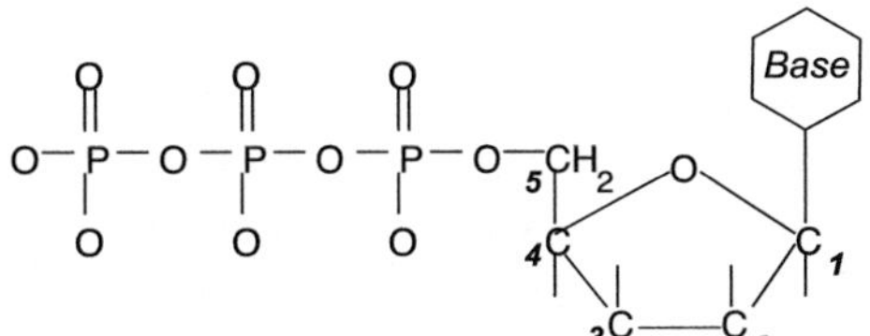

breakup of the polyphosphate group of the nucleotide so the next nucleotide will be attached to the 3' end. The molecule thus always grows in the 5' - 3' direction.

RNA is a single-strand nucleic acid that often folds onto itself, however, DNA occurs in a double stranded form.

A single strand of the DNA grows in the 5' - 3' direction with the addition of nucleotides. As each nucleotide is added, it encounters a slight rotation so the nucleotides end up being

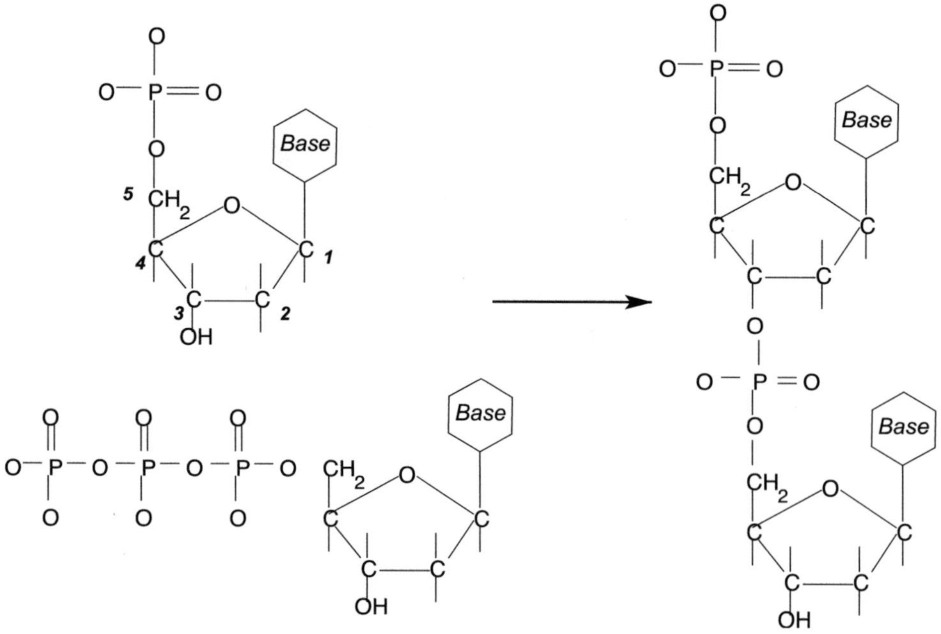

Fig. 2.6 Development of the DNA strand

stacked one on top of the other rotation at each point. These rotations add up to give DNA its characteristic spiral shape. Two strands of DNA join each other through the hydrogen bonding between the bases on each strand as shown in Fig. 2.4. To position the bases next to each other the strands need to be anti-parallel. That is if we head down the double helix in the 5' - 3' direction with respect to one strand it will be the 3' - 5' direction for the other strand.

A DNA molecule that naturally exists in a cell is called a chromosome, which is packaged tightly to occupy less space. We typically see more than one chromosome in eukaryotes and a single chromosome in prokaryotes. The ensemble of the chromosomes of an organism makes up its genome. In eukaryotes, the chromosomes are linear and reside inside the nucleus, coiled around histone proteins. In prokaryotes, the single chromosome is circular and often is folded onto itself for packaging, which is called supercoiling. As prokaryotic cells do not have a nucleus, their DNA is not bounded by a nuclear membrane and resides at an area inside the cell called the nucleoid.

Diploid cells have one pair (paternal and maternal) of each chromosome that make up their genome. Humans, for example, have 23 pairs of chromosomes, except for the sperm and egg cells, which are haploid and contain one of each of the 23 chromosomes. Some cells are polyploid, which contain more than two copies of their chromosomes. Chromosomes are not normally visible except during certain phases of cell division. Other types of DNA seen in organisms are mitochondrial DNA (mtDNA), which consists of a single circular chromosome inside the organelle mitochondria in eukaryotes, chloroplast DNA in plants, and plasmids, which are small non-chromosomal DNA that are often found in prokaryotes and are believed to be helpful but not essential.

2.4   Some History

DNA was discovered long before its significance was known. It was discovered by Fredrich Miescher in 1869 at the University of Tubingen. He named it nuclein because he found it in the nucleus. He had originally isolated it from pus and recognized it to be something new. He later isolated it in much larger quantities from the sperm of salmon. However, he did not connect DNA with heredity. This was done later when Oscar Hertwig at the University of Berlin and Herman Fol in Switzerland discovered fertilization in Salmon and Walther Flemming observed the role of chromosomes in cell division, a discovery that went unremarked until the rediscovery of Mendel's work. Hertwig felt strongly that nuclein was "responsible not only for fertilization but also for the transmission of hereditary characteristics." Miescher did not agree. There was also some problem with the connection between nuclein and chromatin. The chromosomes in the nucleus do not always exist in the condensed form they take on just prior to cell division. Chemical staining easily shows up the chromatin when the chromosome is in the condensed form but not when the chromosomes are in a more relaxed state. It did not make sense that the material responsible for heredity would disappear and reappear. As E.B. Wilson the eminent American biologist said referring to the perceived appearance and disappearance of chromatin "These facts afford conclusive proof that the individuality and genetic continuity of chromosomes does not depend upon a persistence of chromatin" [2]. It was not until the middle of the twentieth century that better techniques finally refuted the idea of the disappearance of chromatin and provided support for Hertwig's position.

The DNA in its sequence of bases contains the information needed to build the molecules that are the workhorses of living organisms—the proteins. The building blocks that are used to construct the proteins are amino acids.

2.4.1   Amino Acids and Proteins

Proteins are made up of a sequence of amino acids. Amino acids are carbon compounds with the structure shown in Fig. 2.7. A Carbon atom called the αCarbon is connected to an amino group (NH_2), a carboxyl group (COOH), a hydrogen atom, and a *residue*. The residues, or side chains, or R groups, can vary significantly in complexity and are responsible for the chemical differences between various amino acids. Based on the structure of the residues the amino acids can be polar or non-polar, aromatic (contain carbon rings) or aliphatic, acidic or basic. Polar amino acids are hydrophilic, that is their residues interact with water, while

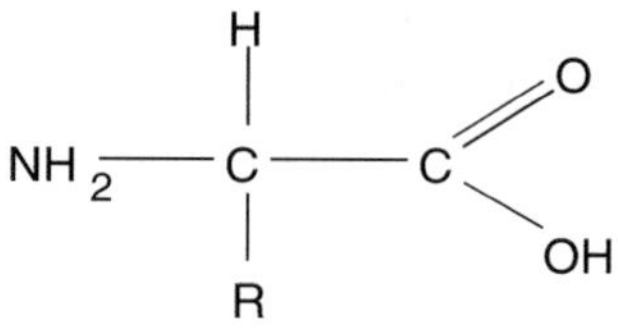

Fig. 2.7 An amino acid

non-polar amino acids are hydrophobic. There are about five hundred amino acids found in nature though only twenty will be of interest to us as these twenty are used by the DNA to form proteins. Some of the amino acids can be synthesized by the body while others cannot be synthesized and have to be ingested. The latter are called essential amino acids (Fig. 2.8).

The reaction of the amino group of one amino acid with the carboxyl group of another results in the formation of a peptide bond. A sequence of amino acids connected via peptide bonds is called a polypeptide. Proteins are complex polypeptides. Just as the DNA molecule can be thought of in terms of a sugar backbone and a sequence of bases, the protein can be thought of in terms of a peptide backbone and a sequence of residues.

Protein is a Greek word meaning "first string" or primary sequence. The name was coined by the Swedish chemist Jons Jakob Berzelius who first discovered proteins in 1838. Proteins are essential to the metabolic and structural functions of life. Enzymes are proteins, as are hormones. Proteins are involved in transport operations in the cell, and they generate the immune response of cells. They form structural elements of the body thus giving us our form. They are the building blocks of life.

Proteins are complex structures. Once the varied amino acids are strung together, because of their properties of polarity or hydrophobic or hydrophilic nature they fold into a three-dimensional shape which is related to their function. Because of the relationship of their organization to their function, a great deal of effort has gone into the study of their organization. The *primary structure* of the protein is simply the sequence of amino acids that makes up the protein. The primary structure is also known as the *covalent structure* as the important bonds here are the covalent peptide bonds that connect one amino acid to the next. The primary structure has two ends, the N-terminal which has a free amino group, and the C-terminal which has the free Carboxyl group. The secondary structure is a local structure induced by hydrogen bonding in the peptide backbone. The most common secondary structures are helical structures called α helices and planer structures called β sheets. The tertiary structure is what results when the entire polypeptide molecule is allowed to fold into its most stable configuration in 3D. Proteins with related functions tend to have similar tertiary structures. Finally, there are proteins that are composed of multiple polypeptide subunits. The configuration of this multi-polypeptide structure is called the quaternary structure of the protein.

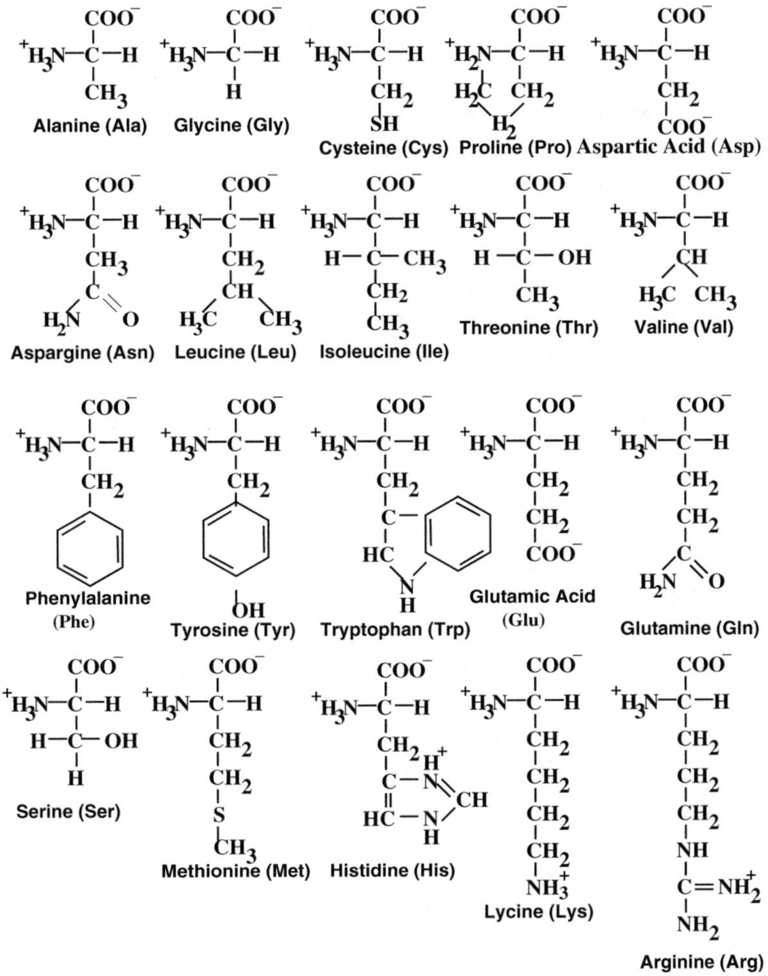

Fig. 2.8 Amino acids [3]

2.4.2 DNA or Protein

Because of its complexity and ubiquity, for a long time proteins were considered to be the most likely candidate for the molecule which mediated the transmission of information from parent to offspring. DNA was considered not as robust as protein and thus less well suited to this important task.

While it was not clear for a long time as to what carried inherited traits from parent to offspring, it was accepted since ancient times that there existed some factor that conveyed the trait from parent to offspring. Modern genetics begins with Gregor Mendel who working with peas conducted a series of precise experiments to determine how traits were passed from parent to offspring. He came up with the idea of recessive and dominant traits and published his results in 1866 in the Proceedings of the Natural History Society of Brunn. Unfortunately, no one paid attention to him. The obscurity lasted until 1900 when three scientists Hugo DeVries, Erich von Tschermak, and Carl Correns [4–6] each independently came up with the same ideas and then independently realized that Mendel had beaten them to it. In 1902 the German scientist Theodor Boveri [7] and an American scientist Walter Sutton [8] independently suggested that the chromosomes (which were thought to mostly consist of proteins) were the bearers of hereditary information. Boveri also suggested that the Mendelian factors existed as distinct entities on the chromosome. Their work was initially resisted and then confirmed by the American scientist T.H. Morgan in 1910. Thomas Morgan began the practice of working with fruit flies (*Drosophila melanogaster*) to study genetics.

The name *gene* for the factor itself was given by Wilhelm Johannsen, a Danish botanist who coined the term in opposition to Darwin's weird and wonderful "Provisional Hypothesis of Pangenesis" which he presented in *The Variation of Animals and Plants Under Domestication.*

While DNA was being dismissed as a candidate for heredity, it was still being studied. In the 1920s the British bacteriologist Fredrick Griffiths was working with two strains of the pneumococcus bacteria; a virulent wild-type fatal to mice and a benign mutant. He found that while heat-killed wild-type pneumococcus did not kill the mice, if the heat-killed wild type were mixed with the mutant benign type, the mice died and the wild-type virulent bacteria was discovered in the dead mouse. Something left over from the dead bacteria was infiltrating the benign bacteria. In the 1940s Oswald Avery, Colin MacLeod, and Maclyn McCarty figured out that this material was DNA. They figured this out by removing one thing at a time from the heat-killed dead bacteria, mixing it with the mutant benign type, and injecting it into mice. After a lot of dead mice, they figured out that removing everything but DNA still killed the mice. To make sure that there was no leftover protein that was causing the transformation of the benign bacteria, they added proteases, which attack protein, to the DNA -mutant bacteria mixture. The mice still died. Only when they added nucleases, which attack nucleic acids, did the death toll of mice decrease thus confirming that it was DNA from the wild type that was transforming the mutant benign type. But was the DNA actually carrying hereditary information?

Evidence for this was generated by another experiment carried out by Alfred Hershey and Martha Chase. They experimented with a virus, the T2 bacteriophage that infects E. coli. The coat of the virus is made of protein, which in turn is made up of amino acids. Two of these amino acids, Cysteine and Methionine contain sulfur. None of the amino acids contain Phosphorus. DNA on the other hand contains no Sulfur but is rich in Phosphorous. Hershey and Chase made up two batches of viruses or phages—viruses that infect bacteria. They allowed one batch of phages to reproduce in the presence of radioactive sulfur (^{35}S) and the other batch they allowed to reproduce in the presence of radioactive Phosphorus (^{32}P). Given that the coat is protein made up of amino acids containing Sulfur, and the DNA does not contain Sulfur but does contain Phosphorous, the radioactive sulfur was incorporated only in the phage shell while the radioactive phosphorus was taken up by the DNA. They then allowed the two sets of phage to infect separate suspensions of E. coli bacteria. The suspensions were then agitated and then centrifuged to separate the bacteria from the viral shells. Checking the bacteria showed that the colony infected by the ^{35}S labeled phage did not exhibit radioactivity while the colony infected by the ^{32}P labeled phage exhibited radioactivity. Thus, it was clear that it was the DNA that had entered the bacteria and not the protein coat. The radioactive bacteria were further cultured and in due course generated more T2 phage thus demonstrating that it was the DNA of the phage and not the protein that was carrying the hereditary information.

2.4.3   The Double Helix Structure

There was sufficient interest in DNA by now that people started working on the structure of DNA. In 1947, Edwin Chargaff found out that the bases Thymine and Adenine occurred in the same quantities as the bases Guanine and Cytosine. By the late forties, there was increasing agreement that DNA was the vehicle for heredity. The question was what was the structure of DNA and could this structure explain the remarkable consistency of heredity? Two groups that were working on this issue were Francis Crick and James Watson in Cambridge and Rosalind Franklin and Maurice Wilkins at Kings College in London. The team of Watson and Crick had been unsuccessfully attempting to build models for DNA and had finally been told to stop working on the problem. Rosalind Franklin at Kings College on the other hand was progressing rather well. She had been using X-ray crystallography to probe the structure of the DNA. She had obtained results that convinced her of several things about the structure. First that it was in the form of a helix, and that the phosphate groups were on the outside. However, she wanted to get some more data before publishing her results. But Franklin faced a different kind of hurdle. Where Watson and Crick had scientific problems, Franklin's problem was of a more personal nature. There was constant friction between her and Maurice Wilkins, also at Kings College about issues of authority. Wilkins, perhaps unknown to Franklin, shared her crystallography data with Watson and Crick during their visit to Kings College. Just prior to this visit, Edwin Chargaff had visited Cambridge and told them about his discovery about the nucleobases. The crystallography data was the final clue and Watson and Crick were the first to publish the structure of DNA. This work led to their winning a Nobel prize in 1962 which they shared with Wilkins. Franklin had passed away in 1958. The Nobel committee at that time did not have a rule against giving posthumous prizes. That rule came into being in 1974.

2.4.4 One Gene One Enzyme Hypothesis

The fact that the structure of the molecule which carried genetic traits was not known did not prevent scientists from trying to mess with these traits. A member of T.H. Morgan's research team, H.J. Muller employed X-rays to induce mutations in fruit flies. His experiments promised to open the doors to the brave new world of genetic manipulation. In the 1940s Wahoo, Nebraska native George Beadle and Alfred Tatum made use of Muller's technology to develop what became known as the *one gene one enzyme hypothesis*.

Working with bread mold (*Neurospora crassa*) which can grow on a minimal diet of sugar, biotin, and inorganic salts they created a set of mutants which would grow on a complete media but not on the minimal media. These mutants, or auxotrophs, were created by irradiating Neurospora with X-rays or UV which is known to harm DNA. They then added various amino acids to the diet until they found the amino acid that when added to the minimal diet allowed the Neurospora to grow. By cross-breeding the mutants with the wild type they found that this peculiarity was inherited in a Mendelian fashion thus showing the genetic basis for the mutations. Furthermore, by creating a bunch of mutants each lacking the capability to produce a single amino acid they could figure out the biochemical steps needed to produce an amino acid.

Let's for the moment skip ahead and assume we know a particular biosynthetic pathway. For example, consider the pathway shown in Fig. 2.9. In this, a precursor molecule is converted to product 1 using enzyme A. Product 1 is converted to Product 2 using enzyme B and Product 2 is converted to Product 3 using enzyme C. Product 3 is essential for the organism to thrive. If the link is broken down anywhere in this pathway preventing the generation of Product 3 the organism will not thrive. Suppose something happens to enzyme C. For this organism to survive we need to provide it with Product 3. If something happens to prevent enzyme B from functioning we can get the organism to thrive by providing it with either Product 3 or Product 2. Because, as enzyme C is still functioning Product 2 will be converted to Product 3. If something happens to enzyme A then we can get the organism to thrive by giving it Product 1 or Product 2 or Product 3. Now consider some results from the Beadle–Tatum experiments. Beadle and Tatum discovered a group of mutants that could not produce arginine. These would grow provided a diet supplemented with arginine. However, some variants could also grow with supplementation with ornithine and some that would also grow with supplementation with citrulline. We can represent the results of this experiment in tabular form as shown in Table 2.2.

From the top and bottom lines of the table, we can see that whatever defect these mutants have results in the non-production of arginine and when supplied with arginine they all flour-

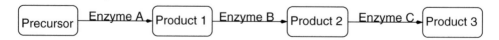

Fig. 2.9 A biosynthetic pathway

Table 2.2 Representation of the results of the Beadle–Tatum experiments

Growth Medium	Mutant 1	Mutant 2	Mutant 3
Minimal	Fails	Fails	Fails
Minimal + Ornithine	Thrives	Fails	Fails
Minimal + Citrulline	Thrives	Thrives	Fails
Minimal + Arginine	Thrives	Thrives	Thrives

ish. Arginine in this case is Product 3. Mutant 1 thrives when given Ornithine, Citrulline, or Arginine, Mutant 2 survives when given Citrulline or Arginine but not when given Ornithine. This implies that Product 1 is Ornithine and Product 2 is Citrulline. Based on many such experiments Beadle and Tatum were not only able to map pathways but they also speculated that each atomic portion of the genome, or gene, was responsible for the production of one enzyme. This became known as the "one gene one enzyme" theory and earned Beadle and Tatum a portion of the 1958 Nobel prize in Physiology or Medicine. More importantly, it opened up a new field of Biochemical Genetics.

2.5   Genetic Code

We have seen two biological sequences DNA, and proteins. The connection between them is through the heritable traits contained in the DNA sequence. The translation from the DNA sequence to proteins is through an intermediate step of messenger RNA formation which we will look at in more detail later. For the moment note that the word "translation" implies that the process of the formation of the primary sequence of proteins can be viewed as a translation from the message written as a sequence of nucleotides to one written as a sequence of amino acids. To write out a message we need a language, and the basic components of a written language are words. Once we start thinking in these terms several questions arise. Are the words fixed length or variable length? If they are of fixed length how long is each word? Are there commas between the words? Is there a grammar that restricts the choices of words in a sequence? To answer each of these questions required the formation of a hypothesis and then the design of experiments to validate or negate the hypothesis. Some of these questions have been answered, others remain open.

Let's assume for the moment that the words in the message are of fixed length. We will come back to this assumption later. How long should these words be? Each word should be able to represent an amino acid and we know that proteins are made up of twenty different amino acids. As there are only four different nucleotides that make up the mRNA the translation is not from a single base to a single amino acid. A wordlength of two bases gives us sixteen possible combinations, or words, which is still less than the required twenty. Therefore, the minimum number of bases per word is three. But with three bases we get sixty-four possibilities which suggests that, either some combinations of bases cannot occur, or that there are multiple codes for each amino acid. Even a cursory examination of DNA and RNA sequences demonstrates that pretty much all combinations of three bases tend to occur in the sequences. Therefore, there have to be multiple three-letter combinations that correspond to the same amino acid.

The elucidation of the code began with the experiments of Nirenberg and Matthaei in 1961. They used the discovery of Marianne-Grunberg Manago and Severo Ochoa, who in 1955 isolated an enzyme called polynucleotide phosphorylase which makes RNA chains without requiring a template. It randomly incorporates whatever bases are around so the final result is a random sequence whose base composition reflects the relative concentrations of nucleotide made available to it. Using this enzyme Nirenberg and Matthaei synthesized a chain of RNA containing only Uracil—a polyU chain. They then set up twenty experiments. In each experiment, the polyU chain was incubated with an *E. coli* extract known to synthesize proteins, all twenty amino acids, as well as other ingredients needed for protein synthesis. In each experiment, a different amino acid was radioactively labeled. The protein formed in each experiment was precipitated and tested for radioactivity. Only the experiment with radioactively labeled phenylalanine contained radioactive protein. The first word of the genetic code had become available; UUU coded for phenylalanine. The same approach then quickly lead to the knowledge that AAA codes for lysine, and CCC codes for proline. The experiment didn't work for GGG. Later it was discovered that polyG RNA chains formed tetraplexes that could not be operated on by the translation machinery.

This approach also works–kind of—for codes other than AAA, CCC, and UUU. Suppose we incubate polynucleotide phosphorylase with a mixture of A and C in the ratio of 5:1. If we randomly selected a nucleotide from this mixture, the probability of getting an A would be 5/6, and the probability of getting a C would be 1/6. From this we can obtain the probabilities of triplets in the RNA sequence made by polynucleotide phosphorylase as follows:

$$\text{Three A's} \qquad \frac{5}{6}^3 = 0.579$$

$$\text{Two A's and a C} \qquad 3 \times \frac{5}{6}^2 \frac{1}{6} = 0.347$$

$$\text{One A and two C's} \qquad 3 \times \frac{5}{6} \frac{1}{6}^2 = 0.069$$

$$\text{Three C's} \qquad \frac{1}{6}^3 = 0.005$$

When we make proteins with this mixture we can look at the fraction of amino acids incorporated in the protein and compare that with the various combinations listed above. Clearly, the amino acid most incorporated, based on our earlier information, would be lysine as the code for lysine is AAA. At least a small proportion would be proline which is coded for by CCC. We say at least because proline could also be coded for by some other triplet. In this experiment, the other amino acids that were incorporated into the protein were asparagine, glutamine, histidine, and threonine [9]. Based on the relative amounts incorporated it was hypothesized that the codewords, or codons, corresponding to Asparagine, glutamine, and threonine contained two A's and a C, while the codes for histidine and proline contained two C's and an A. Similar experiments were conducted using other combinations of bases and the composition (though not the order) of the nucleotides in the codons for the amino acids was obtained. The order was obtained through an experiment developed by Gobind Khorana in 1964. Khorana and his colleagues developed chemical methods for synthesizing RNA with repeating patterns of two or four bases. These were used to specify the final codons. For example, the sequence ACACACACA... resulted in the uptake of equal amounts of threonine and histidine. There are only two possible codons ACA and CAC therefore, one of these must code for threonine and the other for histidine. The earlier experiment had shown that histidine is coded for by a codon containing two C's and an A, therefore, CAC must code for histidine while ACA codes for threonine. In this fashion, experiment by experiment the amino acids corresponding to 61 of the 64 possible codons were discovered. The final three UAA, UGA, and UAG only disrupted the formation of proteins and these were finally understood to be codons that stopped the translation process. Hence they are called *stop* codons, *termination* codons, or nonsense codons. The full code is shown in Table 2.3.

So now that we have a code, how does this code get translated? Francis Crick of Watson and Crick fame suggested that there existed an adaptor molecule that translated the nucleotide sequence to the amino acid sequence. Mahlon Hoagland and Paul Zamecnik, in 1958 showed that the adaptor molecule was a heat-stable soluble RNA molecule which was later characterized by Robert Holley in 1965. This molecule was named transfer RNA, or tRNA. There are a number of tRNA molecules which are specific for different amino acids. In two dimensions they have a clover leaf structure as shown in Fig. 2.10 with a triplet at the point of one of the leaves which base pairs to a codon. Because of its function, this triplet is called the anti-codon. How many different tRNA molecules are there? There has to be a minimum of twenty, as there are twenty amino acids. But are there 61 corresponding to the 61 codons which code for amino acids? As it turns out we can get by with a lot fewer than 61.

If you look at Table 2.3 you will find that there are several codes for many of the amino acids. In fact, except for Methionine and Tryptophan, all other amino acids are represented by multiple codes. If n codes correspond to the same amino acid the code is said to be n-fold degenerate. However, for many of the multiple codes, it seems it is the first two bases that are most important for determining the amino acid. If we know that the first two bases in the codon are CC then, regardless of the third base, the codon codes for proline. Similarly, if the first two bases are AC, regardless of the third base, the codon codes for threonine. Francis Crick suggested that this third base "wobble" was probably due to the mechanistic attachment of the bases. That is, the third base does not attach as strongly as the first two bases and can therefore, attach to more than one base in the anticodon. It turns out that sometimes the position on the anticodon which matches up to the third base on the codon is taken up by inosinate which is a nucleotide containing the base hypoxanthine, and this nucleotide will pair with U, C, or A. Crick proposed that if the base on the anticodon corresponding to the third base on the codon was a C or an A the base pairing was specific.

Table 2.3 RNA Codon table

	U	C	A	G
U	UUU Phenylalanine	UCU Serine	UAU Tyrosine	UGU Cysteine
	UUC Phenylalanine	UCC Serine	UAC Tyrosine	UGC Cysteine
	UUA Leucine	UCA Serine	UAA Stop	UGA Stop
	UUG Leucine	UCG Serine	UAG Stop	UGG Tryptophan
C	CUU Leucine	CCU Proline	CAU Histidine	CGU Arginine
	CUC Leucine	CCC Proline	CAC Histidine	CGC Arginine
	CUA Leucine	CCA Proline	CAA Glutamine	CGA Arginine
	CUG Leucine	CCG Proline	CAG Glutamine	CGG Arginine
A	AUU Isoleucine	ACU Threonine	AAU Asparagine	AGU Serine
	AUC Isoleucine	ACC Threonine	AAC Asparagine	AGC Serine
	AUA Isoleucine	ACA Threonine	AAA Lysine	AGA Arginine
	AUG Methionine Start	ACG Threonine	AAG Lysine	AGG Arginine
G	GUU Valine	GCU Alanine	GAU Aspartic Acid	GGU Glycine
	GUC Valine	GCC Alanine	GAC Aspartic Acid	GGC Glycine
	GUA Valine	GCA Alanine	GAA Glutamic Acid	GGA Glycine
	GUG Valine	GCG Alanine	GAG Glutamic Acid	GGG Glycine

However, if the base on the anticodon was a U or a G the tRNA could bind to two codons. If the base was an I the tRNA could recognize three codons. Using these relationships we end up with thirty-two tRNAs.

The sequential structure of proteins is reflected in the sequential organization of the codes, corresponding to amino acids, in the genes. If we think of the DNA as a long sequence of letters then the genes are subsequences that begin with a start codon and end with a stop codon. The most basic form of organization of information in the DNA sequence is in the form of genes. In prokaryotes generally the entire sequence between the start and stop codons is made up of triplets which code for amino acids. In eukaryotes by contrast not all the sequences between start and stop codons code for genes. There are portions of the sequence that are expressed as amino acids. These portions are called *exons*. The other intervening portions are called *introns*.

Fig. 2.10 Transfer RNA
molecule for the amino acid
alanine

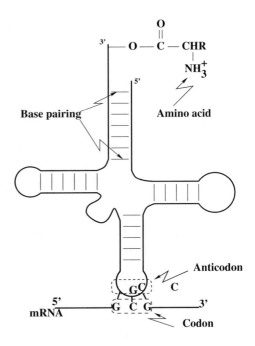

2.6   Summary

So, after all this what do we know about the structure imposed on the DNA by biology
and chemistry? We know that the alphabet that makes up the DNA sequence consists of
four letters, A, C, T, and G. We know that DNA is a directional molecule so the direction
in which the sequence is read makes a difference. We know that the two strands of the
DNA are redundant, in that one of the strands can be obtained from the other by reverse
complementation. We know that there are some parts of the DNA, the genes, which are
functionally different from the other part. The gene has a triplet-based organization because
of the codon structure. The first three letters of the gene are likely to be ATG while the last
three letters are TAA, TAG, or TGA.

DNA is not a static molecule. It participates in a huge number of interactions with the
interactions shaping the composition of the DNA and the composition of the DNA shaping
the interactions. We look at three of the major interactions that DNA participates in the next
chapter.

2.7   Exercises

1. How many nucleotides are there on the human genome (consider only one strand of 23
chromosomes)?

2. What are the three building blocks of a nucleotide?

3. What are the biochemical differences between the DNA and RNA molecules?

4. Write the complementary strand from 5' to 3' direction of the following template DNA sequence

5' C C A T A G G G T C A A C 3'

5. Write down the anticodon on the tRNA molecules that transfer the following amino acids: Asparagine, Serine, Cysteine, Valine.

6. Consider the following mRNA sequence. Write down the corresponding amino acid sequence. Now assume that the "underlined bold C" changes into an A. Write down the corresponding amino acid sequence for this case and explain if this is a critical mutation or not.

5' A U G C C A U A U G G U C A U U A <u>C</u> U G A 3'

7. The NCBI Gene ID for the human "vascular endothelial growth factor A" (VEGFA) gene is 7422. Verify this information on the NCBI website. What is the NCBI Gene ID for the VEGFA gene of Mus musculus, the house mouse?

8. Describe two groupings of nucleotides (e.g., weak/strong) and three groupings of amino acids (e.g., polar/non-polar).

References

1. A. Lehninger. *Principles of Biochemistry*. Worth Publishers Inc., 1982.
2. J. Lederberg. The transformation of genetics by DNA: an anniversary celebration of Avery, Macleod and Mccarty (1944). *Genetics*, 136(2):423–6, 1994.
3. Wikipedia. Amino acid — Wikipedia, the free encyclopedia, 2005. [Online; accessed 1-September-2005].
4. Erich Tschermak. *Über künstliche Kreuzung bei Pisum sativum*. E. Tschermak, 1900.
5. Hugo Devries. Die mutationstheorie. *Ancient Philosophy (Misc)*, 14(a), 1904.
6. Carl Franz Joseph Erich G CORRENS. Mendel's regel uber das verhalten der nachkommenschaft der rassenbastarde. *Ber. dtsch. botanisch. Ges*, 18:158–167, 1900.
7. Theodor Boveri. Uber mehrpolige mitosen als mittel zur analyse des zellkerns. 1902.
8. Walter S Sutton. On the morphology of the chromoso group in brachystola magna. *The Biological Bulletin*, 4(1):24–39, 1902.
9. A. Lehninger, D.L. Nelson, and M.M. Cox. *Principles of Biochemistry, 5th edition*. W.H. Freeman, 2008.

A Bit of Biology

3

In a nutshell .. We met the DNA molecule in the last chapter. Here we see some of the biology that this molecule participates in.

The structure of DNA is the terrain on which bioinformatic analysis operates. This terrain is shaped in many ways by the operations that the DNA molecule participates in. There are three operations that all DNA molecules in living organisms take part in–replication, transcription, and translation—and these operations influence in ways both subtle and not so subtle the organization of the molecule. In this chapter we look at these operations and see if we can trace the impact of these operations on the genome.

3.1   Replication

The most important task of a living organism is replication; witness the amount of energy spent on this task. In the book *The Selfish Gene*, Richard Dawkins makes a plausible case for life itself being simply the result of the effort by genes to replicate and propagate. Given the importance of the task, it is reasonable to assume that replication leaves its imprint on the genome in terms of how information is organized.

Genes are organized linearly along the DNA molecule, and the DNA molecules are organized, for the most part, into chromosomes. In *prokaryotes*, organisms with no organized nucleus, there is usually a single circular chromosome. In *eukaryotes*, which do contain an organized nucleus, the chromosomes are "linear" and reside in the nucleus. Because of the length of the chromosome and the small amount of space available in the nucleus, the chromosomes are packaged in a very complex manner. As an example, the length of the human chromosomes is about two meters while the diameter of the nucleus in which they are contained is only $10\,\mu$m. Therefore, multiple levels of packaging are necessary to fit the chromosomes within the nucleus. At the most basic level of organization, each DNA strand twists 10.4 bases around the helical axis per turn. The number of twists T is defined

K. Sayood and H. H. Otu, *Bioinformatics*, Synthesis Lectures on Biomedical Engineering, https://doi.org/10.1007/978-3-031-20017-5_3

as the total number of turns in the DNA helix. This is simply the total number of bases divided by the number of bases per turn. Fewer or more twists lead to supercoiling. The coiled DNA is wound around proteins called the *core histones* to form *nucleosomes*. The nucleosome containing string is further packaged with the help of other proteins to form the chromosome.

For replication to take place the DNA has to be unpackaged and then repackaged. We will not discuss the packing and unpacking processes except at the most basic level. For now, we will limit ourselves to simply noting the existence of such processes. The basic DNA replication process, at the level at which we are studying it, is similar in both prokaryotes and eukaryotes. Because the process is easier to explain in prokaryotes we will focus our attention on replication in prokaryotes and note the differences with replication in eukaryotes where applicable.

3.1.1   The Most Beautiful Experiment

Each pair of DNA parental strands is replicated to generate two pairs of daughter strands. There are several ways in which this replication could theoretically take place. In *conservative* replication, both strands of DNA would be copied and the result as shown in Fig. 3.1 would be a progeny DNA containing two new strands and the original DNA containing the two original strands. In *semi-conservative* replication, both strands would again be copied but the result would be two double-stranded DNA in which one strand is the old strand and one strand is a new strand (See Fig. 3.2).

Which particular method of replication occurs in practice was determined in 1958 by Meselson and Stahl. Meselson and Stahl raised bacteria in a material enriched with a heavier isotope of Nitrogen, ^{15}N. Recall that nitrogen forms a significant proportion of the structure of the nucleobases. The colony of bacteria that grew in this material incorporated this heavier isotope of Nitrogen in the nucleobases and thus contained a dense form of DNA. The experimenters then transferred the culture to a medium containing normal (^{14}N) nitrogen. Thus, any new DNA that was formed would contain nucleobases with the lighter Nitrogen isotope and would be less dense. Meselson and Stahl sampled the culture at various time

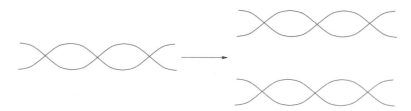

Fig. 3.1 A depiction of conservative replication. We depicted the original strands in *blue* and the new strands in *red*

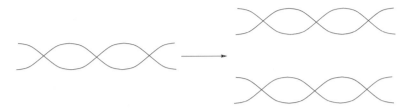

Fig. 3.2 A depiction of semi-conservative replication. The original strands are depicted in *blue* and the new strands in *red*

points and isolated the DNA from these samples. The DNA were then separated using a Cesium density gradient.

If replication was conservative then right after replication the density gradient should contain two bands of DNA, one band corresponding to the parent strands which contained ^{15}N, the heavy isotope of Nitrogen, and one band containing the offspring DNA which incorporated ^{14}N, the lighter isotope of nitrogen. The later samples would contain more and more of the band containing ^{14}N DNA. Instead what Meselson and Stahl found was that the initial samples contained a single band of DNA with density between that of the ^{14}N DNA and the ^{15}N DNA. The later samples contained two bands of DNA, one with a density of the ^{15}N DNA and the other band with density between the ^{14}N DNA and the ^{15}N DNA. This will happen if the replication is semi-conservative. After the first replication, the result will be DNA containing one strand from the parent containing the heavy isotope of Nitrogen and one newly constructed strand which incorporates the lighter isotope of Nitrogen. Therefore, the band from the sample right after the first replication will be between the band formed by the denser DNA containing two strands with the heavy isotope and the band formed by the DNA with both strands containing the lighter isotope. In the next generation, the offsprings will again take one strand from the parent and construct one new strand. The new strand will contain the lighter isotope of Nitrogen while half of the parent strands will contain the heavier isotope and half of the parent strands will contain the lighter isotopes. As the generations progress more and more of the population will contain both strands incorporating the lighter isotope.

The Meselson–Stahl experiment is simple, elegant, and conclusive. All you would ever want an experiment to be. John Cairns, one of the pioneers in the study of replication, called the Meselson–Stahl experiment "the most beautiful experiment in biology." The experiment settled the question of replication as well as removed any doubts about the double helix structure of DNA.

3.1.2   Replication Steps

Replication is a complex operation. A long molecule that has been packaged in complicated ways has to be unpacked and copied with fidelity. DNA has to be replicated with fidelity in order to propagate the organism. In fact, replication is important enough that any problems with the machinery performing replication usually will lead to the death of the cell or organism. As might be expected for such a complicated process there are multiple steps involved in DNA replication. Ignoring some of the unpacking steps, the origin of replication has to be recognized, the DNA strands have to be unwound, separated, kept separated, and copied. Each of these requires a different piece, and sometimes several different pieces, of machinery. And each step leaves its mark on the DNA. Let's go through the various steps of replication.

Initiation

Replication is initiated at one or more specific points called the *origin of replication*. In bacteria, this is generally a single site, while in eukaryotes with much larger linear chromosomes there are usually multiple origins of replication. In bacteria, the region is called *oriC*. Replication initiates at *oriC* and then proceeds in both directions. The *oriC* has been mapped in some detail for *E. coli* in which the *oriC* region is a 245 bp segment containing three sets of repeated subsequences. First on the 5' end is a sequence of length 13 (5'-GATCTNTTNTTTT-3') repeated three times (the sixth and ninth bases are ambiguous, hence the N). This is followed by six repeats of the consensus sequence 5'-AGATCT-3'. The final set is a set of five repeats of the 9 bp sequence 5'-TT(A/T)TNCACA-3'. The (A/T) notation here means that the third base can be either an A or a T. This last set of repeats are called *DNA boxes* and are binding sites for the DnaA protein, also called the replication initiator protein. The binding of the DnaA protein to these sites initiates the unwinding of an AT-rich sequence in the *oriC* region which acts as the loading site for the enzyme Helicase. The fact that this area is AT-rich makes it easier to pull the strands apart as Adenine and Thymine are connected by two hydrogen bonds, as opposed to Guanine and Cytosine which are connected with three hydrogen bonds. We will see this kind of structure in locations of the DNA which need to be pulled apart. The *oriC* region is very important for bacterial survival. Without this region, the bacterial proteins would not be able to replicate the genome.

The unwinding process is initiated by the initiator protein binding to the DNA boxes in the *oriC* region. The initiator protein then recruits the other proteins, such as Helicase and single-stranded DNA binding proteins, necessary for continuing and stabilizing the unwinding. Helicase proceeds to unwind the DNA at a rate ranging from a few turns per second in eukaryotes to several thousand turns per second in prokaryotes. The unwound portion of the DNA is called a *replication bubble* and the points at each end of the bubble are called the replication forks. The replication fork is stabilized by the attachment of *Single Strand Binding Protein*. The actual copying is done by an enzyme called *DNA Polymerase III*

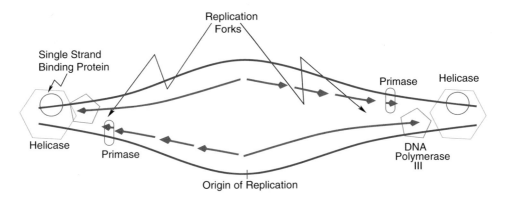

Fig. 3.3 A cartoon representation of a replication bubble

at a rate of up to a thousand bases per second in prokaryotes. However, DNA Polymerase III cannot start copying from scratch. It is only able to add nucleotides to the 3' end of an existing chain. Therefore, another piece of machinery is required to initiate the copying process. This initiation is provided by an enzyme called *RNA Polymerase*, also known as *Primase*. The Primase attaches a short RNA sequence, about 10–12 bp, to the complementary nucleotides on the DNA molecule. This short strand called a *primer* is used by DNA Polymerase III to extend the copy. The primer is later stripped away by an *exonuclease* which is part of DNA Polymerase I. The primer is replaced by DNA polymerase I and the gaps are filled in by *DNA ligase*. Some of the players in this process are depicted in Fig. 3.3.

In bacteria, there is usually only one origin of replication. Eukaryotic chromosomes are a lot bigger and using only a single origin of replication would mean that replication would take a very long time. Consider that the human genome is about a thousand times longer than a typical bacterial chromosome. The use of a single origin of replication would mean a thousandfold increase in the time for replication which is impractical. Thus eukaryotic chromosomes will have multiple origins of replication situated 50–300 thousand bases apart. The human genome has about 30,000 origins of replication.

Elongation

The two strands of DNA are anti-parallel. That means that the 3' to 5' direction on one strand is the 5' to 3' direction on the other strand. As we separate the DNA strands for replication both strands get replicated. However, because of the anti-parallel nature of the strands, they get replicated differently. On the 3' to 5' strand replication proceeds continuously as the strand being built up is built up from the 5' to the 3' direction as shown in Fig. 3.4. On the other strand, however, the original strand is being opened up in the 5' to 3' direction. As the parental DNA is further opened we are faced with an asymmetry between the two strands. On one strand, called the *leading strand* the replication can proceed in the same direction.

Fig. 3.4 Initial stage of
elongation

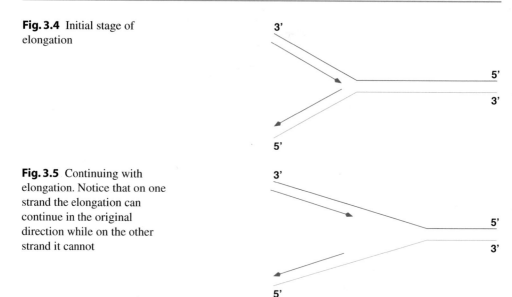

Fig. 3.5 Continuing with
elongation. Notice that on one
strand the elongation can
continue in the original
direction while on the other
strand it cannot

On the other strand, called the *lagging strand* the replication has to be reinitiated as shown
in Fig. 3.5. Therefore replication on the lagging strand cannot proceed continuously and has
to be restarted every so often resulting in the creation of short fragments known as Okazaki
fragments after the person who first discovered them in *E. coli.* .

DNA polymerase III has to do several things. It has to hold onto the DNA. It has to provide
continuous elongation for one strand, and it has to provide discontinuous elongation for the
other strand. In order to complete all these tasks the enzyme has to have a somewhat complex
structure. The polymerase is made up of several subunits. Two subunits form a clamp called
the β clamp. At any given time there are actually three β clamps present. Two are helping
guide the polymerase along the leading and lagging strands, while the third readies to grab
the next primer on the lagging strand (See Fig. 3.6).

Fig. 3.6 Formation of Okazaki
fragments

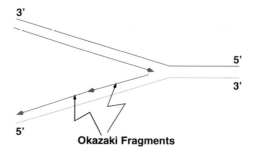

Termination

Termination of replication occurs when replication forks from two different directions meet. The termination process is complex and there are differences between the exact mechanisms in different organisms. The termination mechanism has been most widely studied in *E. coli* so that is what we look at here. In *E. coli* the termination happens in a region opposite the *oriC* region known as the terminus. The replication forks travel at about 1000 nucleotides per second around the circular genome so there needs to be some mechanism to slow down this motion. In the terminus region, there are a number of termination sites known as *Ter* regions made up of the consensus sequence G(T/G)A(T/A)GTTGTAAC(T/G)A. DNA binding proteins encoded by the *tus* gene (which is also located in this region) bind to the *Ter* sites forming *Tus-Ter* complexes that hinder the progression of Helicase. There are ten of these termination sites in *E. coli* organized in two groups of five. Each group slows down the replication from one direction forcing the replication forks to meet in a well-defined region of the chromosome.

Another organism in which the termination mechanism has been studied in some depth is the wonderfully named bacteria *B. subtilis*. In *B. subtilis* the overall mechanism is similar though the sequences that make up the termination sites are completely different.

In eukaryotes with multiple replication forks, the situation is much more complex and varied. Not only are there a large number of replication forks, these forks also have to move through chromatin because of the packaging of the chromosomes. The multiplicity of replication forks and the complexity of the environment precludes the use of the location-dependent replication traps seen in *E. coli* and *B. subtilis*. Instead, termination requires a complex dance between two approaching replication machineries that is not yet entirely clear.

Telomeres

Bacterial chromosomes are circular so each replication process results in an exact replica. Most eukaryotic chromosomes are linear. This causes a problem when replicating the ends of the chromosomes. The polymerase requires a certain amount of sequence in order to "sit" properly. At the ends, there is not sufficient sequence to sit on and the polymerase cannot copy the ends of the chromosome. If there were no remedy the replicated chromosomes would keep getting shorter and shorter. To avoid this, eukaryotes have evolved a strategy involving repeated short sequences called telomeres. The ends of the chromosomes consist of telomeres. An enzyme called *Telomerase* will add telomeres to the end of replicated chromosomes. The enzyme contains its own reference RNA sequence which it uses in the process of creating additional telomeric sequences on the end of chromosomes.

Asymmetry in Bacterial Chromosomes

The replication process induces an asymmetry in the bacterial chromosome. Starting from the origin of replication one half of each strand is replicated, known as the leading strand, (almost) continuously while the other half, known as the lagging strand, is replicated as a sequence of Okazaki fragments. These halves, also known as *replichores* display a number of asymmetries. The manner of replication requires that the lagging strand contain within it a large number of sites to which the primase can attach an initiating primer. If the primase requires, or prefers, a particular sequence of bases to attach to, and we assume that the attachment of the primase is a stochastic event, we would expect to see clusters of the preferred patterns of bases on the lagging strand but not on the leading strand.

We can make use of this asymmetry in the chromosome to develop computational methods for identifying the *oriC* region in bacteria.

3.2   Transcription

Form follows function, therefore, how information is organized in DNA depends to a great extent on the functional responsibilities of DNA. The two most important functions of DNA are replication and gene expression. We have previously looked at what we know about the process of replication and how it affects the organization of information in DNA. We begin now to look at the process of gene expression. The *central dogma* of biology, proclaimed by Francis Crick of Watson and Crick fame in 1958, is that information is transferred from the DNA to RNA to proteins. Crick thought that this path was unidirectional, something we now know not to be true. While most of the time the information flow is from DNA to RNA, the enzyme reverse transcriptase allows RNA to be used as a template for DNA. Regardless, in nature we are, for the most part, dealing with information in the DNA being used to make RNA, which is then used to make proteins. Gene expression is the process through which regions of the DNA are transcribed into RNA molecules, most of which are then processed and translated into proteins. Because the process is somewhat different between prokaryotes and eukaryotes we will look at them separately. We begin by looking at the process in prokaryotes.

3.2.1   Transcription in Prokaryotes

To transcribe or copy a specific section of DNA into RNA we need machinery to recognize the particular section to be encoded; machinery to unwind the DNA; machinery to hold on to the DNA while the copying process takes place; machinery to copy the DNA sequences into RNA and finally machinery to dissociate the RNA from the DNA. All these processes require certain structural characteristics in the DNA sequence being transcribed. At a minimum, there have to be signals in the DNA that indicate where to begin and end transcription. Among the two, the initiation signal has to be, in some sense, stronger because the signal to end transcription has to be conveyed to machinery that is already "reading" the message. The signal to initiate transcription has to "recruit" the machinery for transcription. The signals which recruit the transcription machinery and indicate the location at which transcription is to begin are called *promoters*. The signals that do not actually bind the machinery for transcription but still impact the level of transcription by binding different proteins are called *enhancers* and *silencers*. These signaling sequences, or regulatory sequences, fall into the general category of *cis* elements which are DNA sequences involved in the regulation of gene expression. The proteins that bind to these sequences are called *trans-acting factors*. While regulatory sequences vary among and within organisms we can tease out a general structure for them (with plenty of exceptions).

Let us back up for a minute and establish some terminology that will be helpful as we examine various structures in the genome. The DNA molecule is double stranded and a gene can occur on either of the strands. The primary RNA molecule that results from transcribing the gene will have the same sequence of bases as the other strand with Uracil replacing Thymine. This non-gene strand is referred to as the *coding strand*. The strand containing the gene is referred to as the *non-coding strand* or the *template strand*. As previously mentioned each DNA strand has a directionality. There is a 5' end of the strand and a 3' end of the strand. The gene is transcribed from its 5' end towards its 3' end. From any base, the bases towards the 5' end are referred to as *upstream* bases while the bases towards the 3' end are said to be *downstream* of the base. The first base to be transcribed is called the start point and is numbered +1. The bases downstream from the start point are given consecutive positive numbers. The bases upstream from the start point are given consecutive negative numbers with the first base upstream from the start point being numbered −1. There is no base numbered 0.

With this numbering system in mind let's examine some of the known organization of promoter elements. The two most frequent promoter sequences occur approximately 10 and 35 bases upstream from the start point. As such they are called the −10 and −35 elements. The −10 element has a consensus sequence of TATAAT and is sometimes called the *Pribnow box* after its discoverer. By TATAAT being a consensus sequence we mean that if we examine all such promoters, the first element is most likely to be a T, the second element is most likely to be an A, and so on. In many cases, the promoter for a particular gene will not be identical to the consensus sequence. There are consequences to a particular promoter being very different from the consensus sequence which we will look at later. Some genes have an extended −10 element which consists of TGNTATAAT where N means that position can contain any nucleotide. The −35 consensus sequence is TTGACA. The −10 and −35 elements together are called the core promoter. Some genes also have an upstream promoter (UP) element upstream from the −35 element. Genes that have a UP element are expressed at high levels.

The machinery of transcription, or the proteins which mediate the transcription process, essentially create a nucleotide polymer and are therefore called polymerases. As these polymers are RNA molecules they are called RNA polymerases. There are several different types of RNA but in prokaryotes, all of them are manufactured by the same RNA polymerase. Because the function of the RNA polymerase is different from the DNA polymerase there are significant differences in capabilities and structure. The DNA polymerase creates a lasting template of the genetic material in the organism, therefore, there is a high requirement of fidelity and integrity. The RNA polymerases might create many copies of the same stretch of DNA, some of which might last only seconds and none of which have the "immortal" character of the DNA. While it is important for the functioning of the organism that these copies be true to the original the requirement is not as strict. Therefore the RNA polymerase error rate is several orders of magnitude higher than that of DNA polymerases—an error rate of approximately 10^{-4} compared to an error rate of 10^{-8} for DNA polymerases.

The best-studied prokaryotic RNA polymerase is the *E. coli* RNA polymerase. This polymerase can be split into two parts, the core enzyme, and the σ factor. Together they form the *holoenzyme*. The core enzyme consists of five subunits, namely two α subunits, a β subunit, a β' subunit, and an ω subunit. While the various subunits cooperate in the process of transcription, each seems to have a primary responsibility. The σ subunit is involved with recognizing the initiation point, The β and β' subunits are most involved in the elongation phase while the α subunit is involved in interacting with upstream promoter elements.

The σ factor is responsible for the specificity of the binding of the polymerase. Because, at different times in the organism's life cycle it may need radically different sets of proteins or different amounts of particular proteins, there are a number of different σ subunits each of which recognizes a different promoter. Thus, when the organism is confronted with particular kinds of stress, be it a radical change in temperature or a drastic reduction in the amount of nutrients available, it needs a different set of proteins than it would under conditions of stable growth. The availability of different σ factors which recognize different promoter regions allows the bacteria to respond swiftly to hostile conditions. The σ^{70} subunit which takes part in the transcription of most genes recognizes the canonical promoter with the -10 sequence of TATAAT and the -35 sequence of TTGACA. The σ^{32} subunit which takes part in the transcription of heat shock proteins recognizes a -10 promoter sequence of CCCCAT-TA and a -35 sequence of CCCTTGAA. There are 12 different *E. coli* σ factors.

To provide another level of control on expression the σ factor by itself cannot bind to the promoter. In fact, there is a region in the sigma factor that actively prevents it from binding to DNA. To bind DNA the σ factor needs to be associated with the β' factor of the polymerase.

The process of transcription can be divided into three phases, initiation, elongation, and termination. Let us look at each of these phases and see how the various components of the holoenzyme come into play.

Initiation

Transcription is initiated by the binding of the holoenzyme to the promoter region. The σ factor is the factor in the holoenzyme that recognizes the promoters and is very important in initiating transcription. In experiments with the T4 phage, it was found that the relative transcription activity without the σ factor (i.e. with only the core promoter) was 0.5 as compared to a relative activity of 33 when the σ factor is present. This 66-fold change in activity is a clear indication of the importance of the σ factor. The σ factor improves the ability of the holoenzyme to bind to the core promoters. The binding half-life of the holoenzyme(the time required for half the holoenzymes bound to the DNA to dissociate) bound to the T7 phage is between 30 and 60 h. The same quantity for the core polymerase is less than a *minute*. Clearly, the σ factor makes a huge difference in the ability of the polymerase to bind the promoters. The complex formed by the holoenzyme and the DNA is called the *closed promoter complex*. The tight binding of the holoenzyme to the promoter results in the melting of the DNA at the binding location resulting in the formation of an open bubble. The complex of the holoenzyme with the melted DNA is called the *open promoter complex*. The bubble is more than 10 nucleotides long and is probably more like 17 nucleotides long. This bubble will progress with the holoenzyme during the elongation phase. The holoenzyme then does something odd. While sitting at the promoter it makes several abortive transcripts of length up to 10 nucleotides. Finally, when it has made a stable transcript it leaves the promoter and begins the process of elongation. This action is called promoter clearance. At this point the relationship of the σ factor to the core enzyme changes. Initially, some experiments seemed to indicate that the *sigma* factor leaves the core enzyme and is then recycled. However, results from more recent experiments have left the picture less clear. What is accepted is that after promoter clearance the attachment of the σ factor to the core enzyme becomes less tight.

Elongation

During the initiation phase, the RNA polymerase covers more than 100 bases providing it with a firm grip on the DNA molecule. Once beyond the first ten or so bases the RNA polymerase changes so that it is covering only about 60 bp. This coverage reduces to 30–40 bp after RNA chain extends to 20 bases.

Termination

Termination of transcription can take place in one of two different ways. ρ-dependent termination and ρ independent termination. Each of these leave their own imprint on the DNA.

ρ Dependent Termination

The ρ factor is a hexamer of identical subunits with an RNA binding domain and an ATP hydrolysis domain. The RNA binding domain is necessary for the ρ factor to attach to the RNA transcript being created. The ATP hydrolysis domain is necessary for hydrolyzing ATP to get the propulsive power to move along the RNA. The ρ factor pursues the RNA polymerase, finally catching up when RNA polymerase slows down. This slowing down of the RNA polymerase occurs in the termination region which is a GC-rich region and therefore presumably more difficult to melt than an AT-rich region would be. Once the ρ

factor catches up with the RNA polymerase it unwinds the RNA-DNA hybrid releasing the RNA.

ρ Independent Termination

ρ independent termination occurs in about half of the genes transcribed in *E. coli*. In genes that use ρ independent termination the transcribed RNA has a GC-rich secondary structure near the end of the gene that forms a hairpin. Beyond the hairpin region is a U-rich region. The U-A base-pairing is weak so that when the RNA polymerase pauses after the hairpin the mechanical stresses are sufficient to separate the RNA from the DNA template.

3.2.2   Transcription in Eukaryotes

Transcription in eukaryotes is a more complicated affair than transcription in prokaryotes. To start with there are three different RNA polymerases specialized for different kinds of messages. RNA polymerase I transcribes ribosomal RNA genes to express rRNA, and RNA polymerase II transcribes the RNA for most of the other genes to generate mRNA, as well as some snRNA. RNA polymerase III transcribes tRNA, 5SrRNA and some snRNA. Each of these polymerases functions in slightly different ways. Because most genes are transcribed by RNA polymerase II we focus on it. As was the case with prokaryotes the transcription process can be divided into three phases, initiation, elongation, and termination.

Transcription Initiation for Class II Genes

Unlike the RNA polymerase in prokaryotes, RNA polymerase II does not recognize promoter regions. Instead, the recognition and initial groundwork are done by transcription factors which then recruit the polymerase. While there are a variety of promoter sequences in the neighborhood of the gene, the minimal set required for transcription is called the *core promoter*. For about half the genes the core promoter consists of the TATA box and the initiator (InR) sequence. Other promoter elements include BRE or TFIIB Recognition Element, and DPE or Downstream Promoter Element. The TATA box lies 25 to 30 bp upstream from the initiation site and has a consensus sequence TATAA. The InR is a much more variable sequence with consensus YYCAYYYYY where Y stands for pyrimidine and A is the initiation site.

RNA polymerase II does not bind directly to the promoter sequences. Instead it is recruited to the promoters by various transcription factors. These transcription factors recruit the polymerase, determine the start point and direction of transcription, and set the basal level of transcription. Gene specific transcription factors provide an essential component for the regulation of genes. Class II transcription factors include (in order of appearance) TFIID, TFIIA, TFIIB, TFIIF, TFIIE, and TFIIH.

TFIID contains a saddle-shaped protein—complete with stirrups—called the TATA Box Binding Protein (TBP) which is highly conserved across eukaryotic as well as archaeal species. Contrary to how most DNA binding proteins behave TBP binds along the minor groove of the DNA. This was shown by replacing T and A by C and I (inosine). In terms of electron acceptors and donors, I is very different from A in the major groove but looks like A in the minor groove. Similarly C looks very different from T in the major groove but resembles T in the minor groove. Replacing the T's with C's and the A's with I's did not alter the binding of TFIID to the DNA thus confirming the hypothesis that TBP binds in the minor groove of the DNA.

The TBP binds to the DNA by inserting the "stirrups" between the first T and A of the TATA box and between the seventh and eighth base. This causes the DNA to bend by about $80°$ possibly helping with the melting required to initiate transcription. Along with TBP, TFIID also contains at least eight other proteins called TBP Associated factors or TAFs. In particular the TAFs in TFIID are called TAF_{II}s. The most obvious role of the TAFs is to interact with other promoter elements such as DPE, and InR where they exist. They also bind TBP and thus begin the process of recruiting the polymerase when interacting with TATA-less promoters. Where DPE and Initiators are present $TAF_{II}250$ and $TAF_{II}150$ bind them. In TATA-less promoters with GC boxes, the activator SP1 binds to the GC box and then interacts with the TAFs to tether TFIID. Another important role for the TAFs is to interact with activators during the process of gene regulation. Of the TAF_{II}s the most central seems to be $TAF_{II}250$ which seems to serve as an assembly factor around which the other factors are organized. $TAF_{II}250$ also has two known enzymatic activities. It acts as Histone Acetyltransferase (HAT) and as protein kinase. The TAFs are not necessarily required for the transcription of all Class II genes. In fact certain TAFs are required for less than a fifth of the genes. And to muddy waters further, TBP itself can be replaced by other proteins. Finally, TFIID are heterogeneous in their TAF composition.

TFIID is the first complex binding the promoters. TFIIA then binds, stabilizing the interaction followed by TFIIB. TFIIB is a single polypeptide unit with two distinct domains. One domain interacts with TFIID while the other binds to the TFIIF/Pol II complex. RNA Polymerase II will not bind to TFIID+TFIIA if TFIIB is not present. This provides a method for the regulation of expression of genes. Activators can increase or decrease the expression of a particular gene by enhancing or decreasing the ability of TFIIB to join the preinitiation complex.

The pre-initiation complex is complete with the joining of the factor TFIIH. The complex then unwinds or melts the DNA using the helicase activity of TFIIH and energy from ATP. The complex in this form is called the open Pre-Initiation Complex. The kinase complex in TFIIH then phosphorylates a domain in the largest subunit of RNA polymerase II called the C-terminal domain. The C-terminal domain consists of a sequence of amino acid repeats. The sequence contains Serines and Threonines which get phosphorylated. The phosphorylation of the C-terminal domain brings to an end the initiation phase of transcription.

Capping

After the Pre-mRNA is a few bases long the RNA polymerase pauses for capping. Capping involves attaching a Guanine to the 5' end of the pre-mRNA with a 5' to 5' triphosphate linkages. This 5' to 5' linkage is unusual and therefore not recognized by most endonucleases. This protects the pre-mRNA. Further protection is provided by methylation of the 7 carbon in guanosine and to one or two of the sugar hydroxyl groups of the bases immediately adjacent to the cap.

Along with protection capping also improves the efficiency of translation by about 300 fold. It plays a role in the transport of the pre-mRNA from the nucleus to the cytosol. It facilitates splicing. And it provides attachment to the 40S ribosomal subunit.

Elongation

In the elongation stage, the helicase activity of TFIIH helps TFIIH and RNA polymerase II to move down the genome elongating the RNA. TPB, TFIIA, and other TAFs are left behind to initiate another round of transcription. RNA polymerase II also acquires elongation factors to assist the process. The most important of these is TFIIS. During transcription, RNA polymerases tend to pause at particular sites known as pause sites. The incorporation of TFIIS tends to reduce or eliminate these pauses. TFIIS also seems to stimulate the proofreading of transcripts.

Termination

Unlike the case of prokaryotes, in eukaryotes the RNA polymerase II keeps copying beyond the end of the gene until it passes one or more AATAAA signals. This appears in the pre-mRNA as AAUAAA. An endonuclease recognizes this signal and cleaves the pre-mRNA 11 to 30 nucleotides away on the 3' end.

Cleavage occurs through the agency of two factors Cleavage and Polyadenylation Specificity Factor (CPSF) and Cleavage stimulation Factor (CstF). Both factors are associated with RNA polymerase II. When RNA Polymerase II encodes the AAUAAA message the CPSF attaches itself to this signal. CstF then binds further down at a GU or U rich region following the AAUAAA signal. The DNA then loops possibly assisted by other cleavage factors to bring CPSF and CstF into contact. The loop is then cut through endonucleolytic cleavage leaving the pre-mRNA with a new 3' end. The piece with CstF attached becomes degraded. The poly(A) polymerase which is sitting on phosphorylated CTD begins attaching a poly(A) tail to the 3'end of the pre-mRNA. It attaches about 10 nucleotides and is then stimulated by poly(A) binding protein until the poly(A) tail reaches about 200 nucleotides.

The poly(A) tail provides protection against nucleases though it is not essential for this purpose as witness the transcripts for histone proteins and actin which do not have a poly(A) tail and still survive. However, it does seem to promote stability. The poly(A) tail seems to increase translatability about two-fold. And it has a role in the transport of the pre-mRNA.

3.2.3 RNA Splicing

Unlike prokaryotic genes, many eukaryotic genes contain both coding and noncoding regions. The coding regions are called exons (for expressed regions) and the noncoding regions are called introns (for intervening region). Before the RNA can be translated the introns have to be spliced out. The splicing can take place in a number of different ways including self-splicing, and splicing through the mediation of a spliceosome. The various mechanisms make use of embedded splice signals in the introns. A common example of spice signals can be seen in the following

5'-Exon1-AG/**GU**AAGU–Intron–YNCUR**A**C-Y$_n$N**CAG**/G-Exon2-3'

The bolded nucleotides are invariant—the first invariant A being known as the invariant branch point A.

In the case of spliceosomal splicing, these reactions are mediated by a large RNA-protein complex called the spliceosome. The spliceosome contains five small ribonuclear proteins or snRNP U1, U2, U4, U5, and U6. Each snRNP contains a single snRNA and several proteins. U1 binds to the 5' end of the intron and 3' end of the exon through Watson-Crick base pairing. U2 binds the region around the branch with A not being paired. U4 just complexes with U6 to inhibit it. U5 complexes with U1 and U2 to form a lariat. U6 is activated and displaces U1 and binds U2 to make the catalytic complex.

3.2.3.1 Alternate Splicing

The various alternate splicings of exons can be divided into three groups:

- Group 1: The 5' end of the alternate splicings are different.
- Group 2: The 3' end of the alternate splicings are different.
- Group 3: The 3' and 5' end of the alternate splicings are identical.

Group 1 contains transcripts from different promoter elements. An example is a gene that contains a TATA box both preceding the gene and in the first intron. As the TATA box is a location indicator this means that we can have transcripts that contain the first and last exon. Or one that skips the first exon. These two expressions occur in the salivary gland and liver.

Group 2 contains genes that have multiple polyadenylation sites. An example is the Immunoglobulin heavy chain gene which has two poly(A) sites with the first being a weak site. This means that CstF will bind preferentially to the second (5' side) site. However, if CstF concentrations are higher it will also bind to the first site. In the latter case, the exons between the two sites will not be expressed. These exons code for a transmembrane protein. Thus the first transcript leads to a membrane-bound Immunoglubulin and the second to a secreted Immunoglobulin.

Group 3 contains the skeletal muscle troponin T gene with 64 splice variants. Splicing depends on tissue specific factors.

3.3   Translation

, Once the messenger RNA has been formed we need to translate the message from the RNA to form proteins. This process is called translation. In eukaryotes, translation takes place outside the nucleus in the cytosol so there is a spatial and temporal distance from the transcription process. In prokaryotes, there is no spatial and sometimes no temporal distance between transcription and translation. We can have a messenger RNA being translated while it is being transcribed.

3.3.1   The Process of Translation

The process of translation consists of five steps; activation of amino acids, initiation, elongation, termination, and post-translational modification. Let's briefly look at the first two steps.

Amino Acid Activation

There are two goals to amino acid activation. First, to make sure that the right amino acid gets attached to the right tRNA. Second to provide the energy required to make the peptide bond. The enzymes that do the work of attaching the right amino acids to the right tRNA are called aminoacyl-tRNA synthetases. These synthetases are specific for the amino acid and for the tRNA. The specificity for both is assured in different ways for the amino acid and for the tRNA.

Recall that the amino acid has the form (See Fig. 3.7) where R is the group that identifies the amino acid. The tRNAs are 73–93 bp long RNA molecules that are folded into an L-shaped three-dimensional structure. The two-dimensional structure looks like a cloverleaf with three "leafy" arms and an open arm. An example of a tRNA is shown in Fig. **??**. The two arms of interest to us are the anticodon arm, which contains the anticodon, and the open amino acid arm to which the amino acid is attached. The anticodon leaf always has seven unpaired nucleotides. The amino acid arm is the 3' end of the tRNA. The bases at the 3' end are always 3'ACC. Therefore, the amino acid gets attached to the terminal A.

The attachment of the amino acid to the tRNA is a multistep process involving the aminoacyl-tRNA synthetase enzyme. The aminoacyl synthetases are very specific in terms of the amino acid they bind to the tRNA. The specificity is enforced by a two-stage process. First the enzyme rejects amino acids that are bigger than the correct amino acid for the

Fig. 3.7 Amino acid

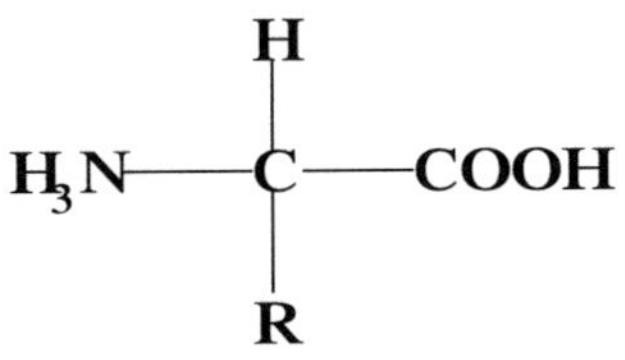

$$H_3N \quad \overset{\displaystyle H}{\underset{\displaystyle R}{\mid}}\!\!\!\overset{\displaystyle \mid}{\underset{\displaystyle \mid}{C}}\!\!\!\overset{\displaystyle \mid}{\underset{\displaystyle \mid}{}}\quad COOH$$

particular synthetase. Second, most of the synthetases also have hydrolytic sites which destroy activated intermediates that are smaller than the correct species.

The enzyme is also specific about the tRNA it binds. The specificity comes from recognizing specific nucleotides, most concentrated on the anticodon arm and the amino acid arm, on the tRNA. The set of nucleotides that allow the aminoacyl-synthetases to recognize specific tRNA is sometimes referred to as a second genetic code.

Once the amino acid binds to the tRNA the specificity of where the tRNA binds is dictated entirely by the anticodon. This was demonstrated by an experiment in which a tRNA was first charged with cysteine. The cysteine was then converted to alanine in a process using Raney nickel. The modified tRNA deposited the alanine to the location where the cysteine was to have been placed.

Ribosomes

The next three stages of translation; initiation, elongation, and termination, are carried out on the ribosome. The ribosomes of prokaryotic and eukaryotic organisms, while they are broadly similar, differ in the details of their structure. They both consist of two major subunits, called the small subunit and the large subunit. In both cases, the small and the large subunit stay separate until the process of translation has been initiated. The sizes of the various components of the ribosome are expressed in terms of Svedberg units (S) which correspond to the sedimentation rate of a particle during ultracentrifugation. The Svedberg measure is not additive. For example, in the case of prokaryotes the small subunit is a 30S unit and the large subunit is a 50S unit but the ribosome is 70S rather than 80S. In prokaryotes the small subunit consists of a number of proteins and the 16S ribosomal RNA (rRNA). The large subunit or the 50S subunit consists of a number of proteins and the 23S and 5S rRNAs. In eukaryotes, the small subunit or 40S subunit consists of proteins and the 18S rRNA. The large subunit or 60S consists of proteins, the 28S, the 5S, and the 5.8S rRNA.

There are three sites of particular interest in the ribosome called the Exit or E site, the Peptidyl or P site, and the Aminacyl or A site. The ribosome provides the environment for the tRNAs to recognize the codons and for the amino acids to form peptide bonds. The ribosome moves along the mRNA one codon at a time in the 5' to 3' direction. It accommodates two aminoacyl tRNAs at a time, one in the P site and one in the A site.

Initiation in Prokaryotes

Initiation is the rate-limiting step in the translation process. In prokaryotes, it requires two cis acting sequences, the Shine-Dalgarno sequence, and the start codon, and three trans acting factors, IF-1, IF-2 and IF-3. The initiation takes place on the small subunit so it is necessary for the small subunit to separate from the large subunit. The factor IF-1 helps this to happen. Once the two units are separated the initiation factor IF-3 binds to the small subunit and keeps the subunits apart. IF-1 occupies the A site preventing the initiation codon from binding there.

The 16S rRNA recognizes the Shine-Dalgarno sequence and positions the mRNA such that the start codon (AUG) is located at the P site. The Shine-Dalgarno sequence was discovered by John Shine and Lynn Dalgarno in 1975. It is a purine-rich sequence with consensus AGGAGG located 8 to 13 bases upstream of the start codon AUG in the messenger RNA. The 16S rRNA in prokaryotes has the complementary sequence (also known as the anti-Shine-Dalgarno sequence) at its 3' end. While AUG is the start codon in most locations in some cases GUG or UUG act as the start codon. When the small 30S subunit of the ribosome encounters the messenger RNA, the pairing of the Shine-Dalgarno and anti-Shine-Dalgarno sequences position the mRNA in such a way that the start codon is located at the P site of the 30S subunit. The triplet AUG codes for methionine. However, in prokaryotes, when it is the first codon at the P site it refuses to accept the tRNA for methionine (tRNA^{Met}). Instead it will only accept a special kind of tRNA that binds N-formylmethionine (fMet). In this amino acid hydrogen is replaced by a formyl group at the amino terminal of methionine. The tRNA that carries this N-formylmethionine is different from the tRNA that carries regular methionine and is denoted by tRNA^{fMet}. tRNA^{fMet} recognizes GUG and UUG as well as AUG. Once the polypeptide chain is about fifteen amino acids long often the formyl group or the methionine is removed. The formyl group is removed by deformylase, the methionine by aminopeptidase (See Fig. 3.8).

The initiation factor IF2 with bound GTP and the initiating fMet-tRNA^{fMet} bind to the complex of 30S, mRNA, IF-1, and IF-3. GTP bound IF2 delivers the fMet-tRNA^{fMet} to the P site of the 30S unit. Remember that the A site is being blocked by IF-1. The anticodon loop of the fMet-tRNA^{fMet} binds to the initiating AUG (or GUG or UUG) codon on the messenger RNA. Hydrolysis of GTP results in the dissociation of the initiation factors and the 50S subunit joins to form the initiation complex.

Fig. 3.8 N-formylmethionine

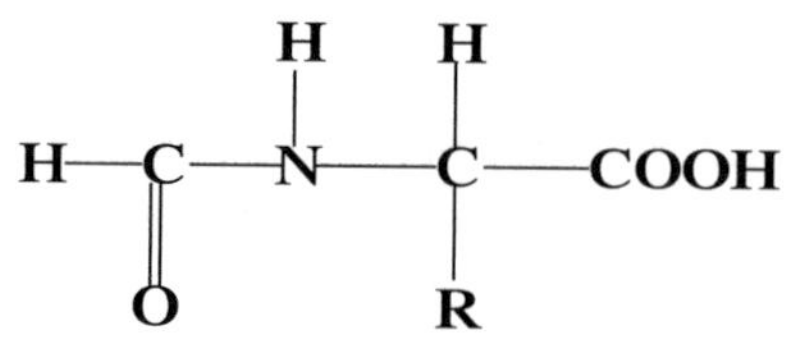

Initiation in Eukaryotes

Initiation in eukaryotes takes place on the small subunit, a 40S unit in eukaryotes, of the ribosome. The eukaryotic cells have at least nine initiation factors. A complex eIF4F which includes three initiation factors binds the 5' cap while the polyA binding protein (PAB) binds the polyA terminal of the messenger RNA.

Separately, eIF2 forms a ternary complex with GTP and the initiator tRNA. The initiator tRNA Met-tRNA$_i^{Met}$ is not formylated but it is a specific tRNA used only for initiation (hence the subscript i). This ternary complex associates with the 40S subunit to form a pre-initiation complex.

The start AUG codon is detected not by the Shine-Dalgarno sequence but by a process of scanning which begins at the 5' cap and is assisted by the helicase activity of several of the initiation factors. The AUG thus found is not necessarily the first AUG but the AUG in a particular context. The best context of the AUG was discovered by Marilyn Kozak to be ACCAUGG and is known as the Kozak consensus sequence.

3.4   Applying Our Knowledge of Biology

Understanding biological processes allows us to make use of this understanding to develop applications that help us extract information contained in biological molecules. Let's briefly look at a very few of the applications.

3.4.1   Artificial Replication—The Polymerase Chain Reaction

We often want to make copies of particular sections of the DNA for further study. This might be because we are looking for distinguishing patterns of mutations. When we speak of DNA, we talk of human DNA or mouse DNA, as if all organisms belonging to a species have the same DNA. In a broad sense, this is true, however, "same" here does not mean identical. Two human beings of the same gender can differ in millions of locations along the genome. Many of these differences are single bases which are different for different individuals. These are called *single nucleotide polymorphisms* or SNPs (pronounced snips). Groups of SNPs that are statistically dependent form *haplotypes* that may correspond to ethnicity or that may be statistically associated with certain types of disease. There are patterns of repeats in human DNA that seem to be specific to individuals. These often go under the title of DNA fingerprint. Rather than sequence the entire chromosome to check these fingerprints, it would be useful to have a method to selectively make multiple copies, or amplify, this portion of the chromosome. Bacteria do not have too many visually discriminatory characteristics. However, some of them can kill us while many of them help us survive. One way to distinguish between them is to look at particular genes on their chromosomes. It has been found that certain genes such as the 16S ribosomal RNA genes have discriminatory characteristics. It would be useful to be able to just replicate the 16S genes in a bacterial DNA sample to figure out the species of bacteria present in the sample.

Artificially replicating DNA is called DNA amplification. There are several ways DNA can be amplified. One way is to incorporate the piece of DNA that we want to amplify into the genomic material of bacteria and then let the bacteria proliferate. Once the bacteria has reached a sufficiently large population we can extract the DNA of interest from the bacteria. The DNA is usually inserted into *plasmids* in the bacteria which are small circular DNA molecules separate from the bacterial chromosome. Amplification can also be performed using yeast cells.

Another method of replication, that does not use living organisms like bacteria or yeast, is based on the fact that DNA strands can be separated, or denatured, or melted by heating. When the DNA is cooled down the strands of the DNA attempt to get back together. This fact is used in a process called the *Polymerase Chain Reaction* or PCR. This process was developed by Kary Mullis and Cetus corporation in 1983 for making multiple copies of, or amplifying, DNA strands. Mullis received a Nobel Prize for his efforts in 1993. The idea is very simple. Heat the DNA in solution together with primers for specific regions on the DNA strand, and nucleotides that will make up the copied strand. The primers are complementary to each strand upstream from the region to be amplified. Once the DNA melts, cool the solution down so that the primers anneal to the DNA. Add DNA polymerase. This will latch on to the primers and start making copies in the 5' to 3' direction on each strand. After giving the process enough time for the polymerase to copy the region of interest, heat the solution again to separate the DNA strands and their copies. We now have twice as many DNA strands as before. Unlike the primers in nature, the primers used in this process are DNA primers rather than RNA primers so we have exact copies of the DNA.

In the initial version of PCR at this point, we would have to add more DNA polymerase because most DNA polymerase is destroyed at the temperature required for melting DNA. If we have DNA polymerase available the procedure can be repeated to double the amount of DNA again. However, the need for adding polymerase slows down the process. In modern day PCR the DNA polymerase used is from the thermophilic bacteria *Thermus aquaticus*. These bacteria exist in the very high temperature of hot springs. As with all living organisms *T. aquaticus* also uses DNA polymerase for replication. However, because of its living environment, the DNA polymerase from *T. aquaticus* can survive high temperatures. This polymerase known as *Taq* polymerase is what is used in modern-day PCR.

The steps used in the PCR process are as follows:

1. Heat to about 96 °C to melt.
2. Lower temperature for annealing (45–60 °C).
3. Elongation to let the DNA Polymerase work. 1 min per kbp. About 70–75 °C.
4. Repeat.

The number of copies of DNA grows exponentially before reaching a plateau (See Fig. 3.9).

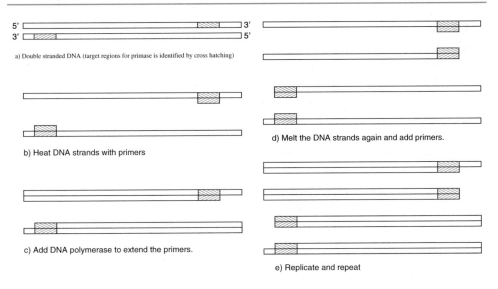

a) Double stranded DNA (target regions for primase is identified by cross hatching)

b) Heat DNA strands with primers

c) Add DNA polymerase to extend the primers.

d) Melt the DNA strands again and add primers.

e) Replicate and repeat

Fig. 3.9 The polymerase chain reaction process

The discovery of PCR had a major impact on the development of molecular biology. PCR amplification allowed for the detection of minute quantities of one particular DNA sequence in a mixture of DNA. This has had applications in forensics for genetic fingerprinting, in genetic testing for the detection of genetic mutations, and in diagnostic testing for viral and bacterial infections. Quantitative PCR is used to monitor the progress of gene expression and the use of PCR with 16S ribosomal RNA genes has revolutionized the study of microbiology.

Quantitative PCR, also known as real time PCR or RT-PCR uses dyes of fluorophores to monitor the progress of the PCR reaction.

The design of primers for use in PCR reactions is a common 'bioinformatic' task. The primer is usually around 21 bases long. The primer has several properties:

1. The GC content is around 50%.
2. The melting temperature should be around 60 °C. The melting temperature can be computed in a number of different ways. An approximate method is $2 \times$ number of A's and T's $+ 4 \times$ number of G's and C's.
3. There should not be runs of three or more C's and G's at the 3' end.
4. The primer should not be self complimentary to avoid hairpins.
5. The forward and reverse primers should not be complementary (especially at the ends) in order to avoid dimerization.

3.4.2 Gene Finding

In many biological applications in medicine, agriculture, and food production we often need to identify the location of genes on a sequenced genome. We already know that genes begin with a start codon and end with a stop codon. However, not every sequence of DNA which begins with a start codon and ends with a stop codon is a gene. These combinations can also occur randomly. Sequences that lie within a start codon and a stop codon are called *open reading frames* (ORFs). Once an ORF has been identified we still need to determine whether it is a gene. Computational methods to determine this use the fact that, in nature, for transcription to proceed the machinery for transcriptions also has to "find" the gene. Transcription like all activities leaves its imprint on the genome. Because the activity involves genes it imposes structure onto the genes. We can use this structure to determine whether an ORF we have identified is a gene. In the case of prokaryotes, we can look upstream from the ORF to see if we can identify promoter regions. We can also look downstream from the ORF to see if we can identify termination signals for ρ dependent or ρ independent termination. For the latter, we can look for a GC-rich region which would form a hairpin followed by an AT-rich region.

The translation process in prokaryotes requires the existence of the Shine-Dalgarno sequence upstream from the gene. Thus the presence or absence of this sequence can be used to evaluate the possibility of a particular ORF being a gene. In the eukaryotic candidate, the Kozak sequence can play a similar role.

In eukaryotes, the promoters are more varied and hence more difficult to use in order to verify whether a particular open reading frame is a gene. However, the polyadenylation signal is more prevalent and can be used as evidence in testing a candidate for a gene. The presence of splice sites can also be used in eukaryotic gene finders to determine whether a particular sequence is a gene.

3.4.3 Finding *OriC*

There are a number of ways we can use our knowledge of the biology of replication to find the origin of replication in bacteria.

The dnaA protein is a replication initiation factor that binds to a 9-base long sequence introducing bending in the DNA strand [1]. This then leads to an unwinding of an AT-rich region in oriC which allows the binding of the replicative helicase. The 9-base long sequence is called a dnaA box. In E. coli the dnaA box is *5' - TTATCCACA - 3*. In other bacteria, and even archaea, similar dnaA boxes can be found. As the function of the dnaA box is to bind the replication initiation factor it is reasonable to assume that the dnaA boxes will be most numerous in the oriC region. One approach to searching for the origin of replication is to look for regions containing multiple dnaA boxes.

When we discussed replication we noted the asymmetry in the process where on the leading strand the primase only needs to bind once to the DNA and replicate that strand in a continuous fashion while on the lagging strand the replication has to be reinitiated multiple times resulting in Okazaki fragments. This asymmetric process also leads to an asymmetry in the distribution of sequences of nucleotides. By looking at these asymmetries we can identify the origin of replication (*oriC*) in bacteria.

The interaction of the DNA molecule with its environment is in part mechanistic and in part stochastic. The latter is necessary if the functioning of this process is not to become prohibitively complex. The mechanistic aspects of the DNA process are reflected in the double helix structure of the DNA, the base pairing rules, among others. By its very nature, the effects of the stochastic interactions are less definite, less striking, and less evident than the structural effects of the mechanistic interactions. However, they have to be present for efficient functioning.

Certain interactions, such as replication, can only be initiated if a complex molecule selectively binds to a particular sequence of bases. Absent a mechanism by which a sequence of bases would attract a particular molecule, other than providing it a convenient binding site should it be in the neighborhood, we have to assume that the binding of the molecule to the recognition site is a stochastic process. If this process is important to the survival of the organism it is reasonable to expect that there will be evolutionary pressure to make the binding more likely. The existence of multiple recognition sites in the neighborhood would increase the probability that binding would occur. Therefore, there is an evolutionary pressure for the clustering of these recognition sequences in the appropriate regions. Clearly, this tendency would be modulated by having to contend with other evolutionary pressures.

This means we can look for "important" sequences by looking at their relative clustering. We define a cluster to exist if the number of that particular sequence in a window exceeds a threshold. We can count the number of clusters in the genome for a particular codon and by comparing this to the number of clusters of other codons, we can make a decision about the relative level of clustering. As an example, we visualize the relative number of clusters of trinucleotides in the *E. coli* genome in Fig. 3.10. We can see from the Figure that there are two sets of triplets that show a high degree of clustering, namely CTG and AGC. The trinucleotide CTG is the preferred binding site for primase in *E. coli*. By now examining how this trinucleotide clusters we can get an idea about the location of *oriC*.

The DNA sequence is not simply a blueprint. It is an active participant in the life of the cell and the organism. As such, genomic sequences are not simply a bag of genes. The chromosomes containing the genes are shaped by the various processes they take part in. Understanding how the DNA is shaped by these processes can help us be more effective and more efficient in carrying out the computational tasks we may wish to perform.

In the next chapter, we begin looking at one of the most common bioinformatic tasks - the alignment of sequences which is a prerequisite to determining similarities and differ-

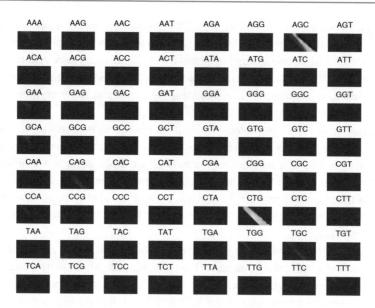

Fig. 3.10 Clustering of triplets in *E. coli*. Each panel represents the relative clustering of a particular triplet as a function of window size and threshold. Low clustering shows up as *blue* while higher levels of clustering are represented by colors on the *red* end of the spectrum

ences between sequences, and parts of sequences. This ability to compute similarities and differences are used in a wide variety of applications from phylogenetic applications to gene function identification.

3.5   Exercises

1. Describe three differences between prokaryotic and eukaryotic cells.
2. Given the following DNA sequence, write down the corresponding mRNA and amino acid sequences. Now assume the "underlined bold C" changes into a T. Write down the corresponding mRNA and amino acid sequences for this case and explain if this is a critical mutation or not.

DNA Template Strand: 3′ T A C T T C A A A <u>**C**</u> C G A T T 5′

5′ A T G A A G T T T G G C T A A 3′

3. Describe promoter, transcription factor, transcription initiation complex.
4. Explain for eukaryotes how RNA splicing and RNA processing work.
5. Locate the Human VEGFA gene in the UCSC genome browser. Which chromosome is this gene on? How long of a genomic region does this gene span? Is this gene on the p or the q arm of its chromosome?

6. Continue from Q5 by considering the region that spans 500K bp upstream (to the left) of the Human VEGFA gene. How many RefSeq genes are there in this region?

7. Using the BioMart tool in Ensembl, find out the GC% of the human genes VEGFA, VEGFB, and VEGFC. Suggestion: On the BioMart page, choose "Ensembl Genes" and then "Human Genes" as your dataset. Under Filters → Gene, use Gene Names to define your input and under Attributes → Gene, pick GC% among other identifiers.

8. Using the UCSC genome browser, identify the number of SNPs on the Human VEGFA gene based on the latest common dbSNP version.

Reference

1. C. Speck and W. Messer. Mechanism of origin unwinding: sequential binding of dnaa to double and single-stranded dna. *EMBO Journal*, 20(6):1469–1476, 2001.

Pairwise Sequence Alignment

4

In a nutshell. A lot of bioinformatics consists of evaluating the similarity and differences between molecules, be they protein, RNA, or DNA molecules. Our understanding of the similarity of two sequences, be they amino acids or nucleotides, is based on how we line up the amino acids or nucleotides in one molecule with the amino acids or nucleotides in the other molecule. We begin by looking at how we align two sequences.

4.1   Introduction

Given two words BARRAYAR and ARRAYAR we can easily tell that the only difference between the two is that the first letter in the second word has been deleted. If we examine this statement more closely we find that in order to reach this conclusion we first *aligned* the sequences as:

$$\text{B A R R A Y A R}$$
$$\text{– A R R A Y A R}$$

Notice that in order to align these two sequences in this manner we had to insert a space before the second sequence. If we had not aligned the sequence in this manner we could have compared the sequences letter by letter and concluded that the two sequences differed in all but one place.

$$\text{B A R R A Y A R}$$
$$\text{A R R A Y A R –}$$

We could argue that the latter comparison is more valid because we did not do anything "artificial" like inserting a space prior to the second sequence. In fact, though, in this second comparison, we have also inserted a space, only at the end of the second sequence. Whether this is any less artificial than the other case is arguable. In the language of sequence alignment, a space is called a *gap*. In order to align the sequences BARRAYAR and ARRAYAR we

K. Sayood and H. H. Otu, *Bioinformatics*, Synthesis Lectures on Biomedical Engineering, https://doi.org/10.1007/978-3-031-20017-5_4

inserted a gap into the second sequence. We can also say that we aligned the first B with a gap. We insert gaps in sequences so that the lengths become the same. We would not have a gap aligned with a gap. If that were the case we could just delete the gap in both sequences. A more formal way of saying this is to say that no column of an alignment should consist of only gaps.

If we wish to compare sequences in any meaningful manner we need to first align them. How we would align two sequences is clear when we have two sequences like BARRAYAR and ARRAYAR which are very similar to each other. However suppose we would like to compare ARRAKIS with BARRAYAR and BARRACK, and make a judgement as to whether ARRAKIS and BARRAYAR are closer to each other or whether ARRAKIS is closer to BARRACK. In each case we have a multiplicity of alignments to choose from. This is clearer when we compare ARRAKIS to BARRACK. Here are two possible alignments:

$$\text{B A R R A C K} -$$
$$\text{A R R A K I S}$$

or

$$\text{B A R R A C K} - -$$
$$\text{A R R A} - \text{K I S}$$

We can show that if we make sure that we do not include alignments in which a gap is aligned with a gap, the number of possible alignments of two sequences of lengths m and n is given by

$$\frac{(n + m)!}{n!m!}$$

So which of these two alignments out of all possible alignments is better? Is it better to insert a space in the middle and get six letters that are the same in both sequences, or is it better to only have five letters that are the same and not insert "artificial" spaces? We cannot answer that question without the notion of scoring an alignment. Scoring an alignment means assigning a numerical score to the alignment which reflects the quality of the alignment; a higher score means a better alignment. We take a look at this in Sect. 4.3. Once we know how to score an alignment we will still have to actually perform the alignment. In our toy example, this was easy. When we are dealing with sequences that are thousands or millions of bases long we will need a more systematic approach. We will look at some options later in this chapter. Finally, an alignment is in essence a hypothesis that one sequence arose from another through a process of mutation—that the sequences are homologous. To estimate the likelihood of the two sequences being homologous we need to find the statistical significance of the alignment score. We will look at this in the last section of this chapter.

4.2   Dot Plot

A very useful method to get an initial idea of the similarities between two sequences is the dot plot. This is a matrix in which the sequences being compared are written along the top and side of a matrix. For each location in the matrix we put a dot where the corresponding elements of the sequences are the same. Let's see how this is done using a few examples. First, let us look at the dot plot for some toy examples. We will compare BARRAYAR and ARRAKIS and TUESDAY and FRIDAY.

By looking for diagonal lines in the dot plot matrix we can see the regions of alignment. The diagonals that descend from left to right show the regions which align when the two sequences are written as ARRAKIS and BARRAYAR. The diagonals that ascend from left to right show regions that align when we reverse one of the sequences. In genomes, this would indicate rearrangement of that portion of the genomic sequence.

From Fig. 4.1 we can see that the pattern AR occurs once in the sequence ARRAKIS and twice in the sequence BARRAYAR, once towards the beginning of the sequences, and once at the end of BARRAYAR. We can see that the pattern ARRA occurs in both sequences. This particular region would also align if we reversed this region in one of the sequences. Furthermore, we can see that all the alignments occur at the beginning of the ARRAKIS sequence. If we look at Fig. 4.2 we can see that there is only one region of alignment and this is at the end of the two sequences. A final note; in these dot plots, we have written one of the sequences in each pair, ARRAKIS in the first case and FRIDAY in the second from top to bottom. As we shall see, sometimes these are written from bottom to top. Both ways of writing the sequences are equally informative.

We really didn't need the dot plot to tell us the similarities between these sequences; it is easy to pick out the similarities just by looking at them. However, when we compare genomes that are much longer, alignments are difficult to pick out without the help of something like a dot plot.

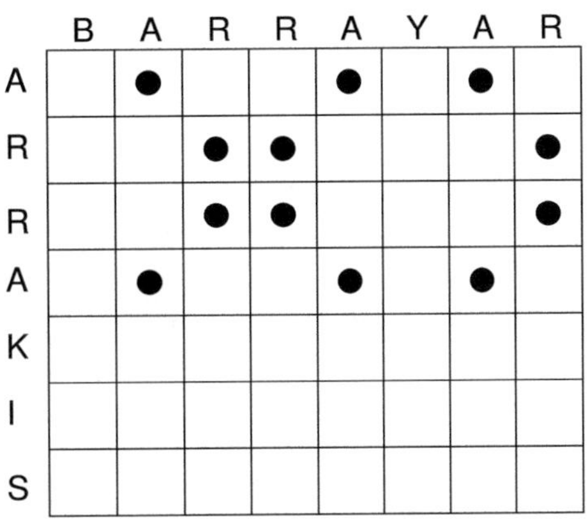

Fig. 4.1 Similarity between BARRAYAR and ARRAKIS

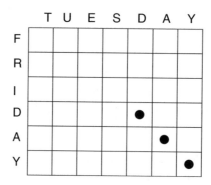

Fig. 4.2 Similarity between FRIDAY and TUESDAY

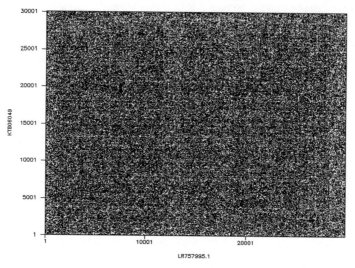

Fig. 4.3 Dot plot of sequences from MERS and SARS Cov2 viruses with a blocksize of 6

When using a dot plot for DNA sequences we encounter a different problem. With only four letters in a sequence that is on the order of hundreds of thousands or millions of bases long, the number of matches for each letter are going to be enormous. If we plot these matches all we will see is a sea of dots. In order for the dot plot to be more informative we group bases together and generate a plot where the groups of bases are the same. Consider Fig. 4.3. This is the dot plot obtained when comparing the sequence of a MERS (middle east respiratory syndrome) virus isolated in Saudi Arabia and the sequence of a Covid19 virus isolated in Wuhan, China. The MERS sequence is written from bottom to top in this plot. You can see that despite using a sliding window of 6 bases to create a sequence from an alphabet size of 4^6 we still have a very noisy dot plot. If instead, we use a sliding window of ten bases

Dottup: fasta::emboss·dottup−E20210316−111536−0859−98430...
Tue 16 Mar 2021 11:15:39

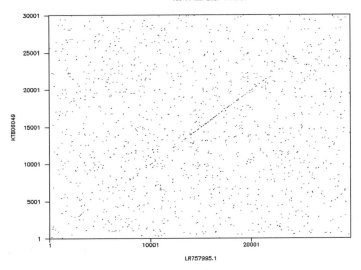

Fig. 4.4 Dot plot of sequences from MERS and SARS Cov2 viruses with a blocksize of 10

Dottup: fasta::emboss·dottup−E20210318−084148−0451−25654...
Thu 18 Mar 2021 08:43:27

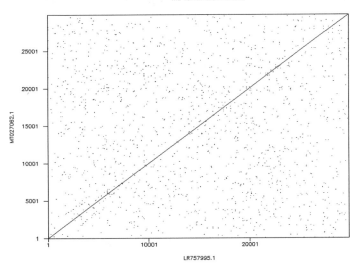

Fig. 4.5 Dot plot of sequences from SARS Cov2 viruses from California and Wuhan with a blocksize of 10

together we get the somewhat more informative dot plot shown in Fig. 4.4 which identifies regions of similarity between about location 14,000 and 20,000 in the MERS sequence.

If we now compare two SARS Cov2 viruses, one from California and one from Wuhan, as shown in Fig. 4.5 we can see the expected alignment throughout the genomes. Clearly the two sequences are very similar. (All of these figures were generated using the Dottup software on the EMBOSS website https://www.ebi.ac.uk/Tools/seqstats/emboss_dottup/.)

4.3   Scoring Alignments

When we align two sequences we usually have multiple choices of possible alignments and we need some way of deciding which one to choose. Usually, when we are trying to decide if one thing is better than another we have to define an objective function or a cost function. The objective function measures how well our objective is satisfied by the different choices. The cost function measures the cost associated with each choice.

In order to define the objective function let us step back and look at why we are measuring the distance between sequences. Broadly speaking we are attempting to find the degree of the evolutionary relationship between the two sequences. There is an implicit assumption that the two sequences are somehow evolutionarily related. One way they could be related is if one arose from the other. In this setting, we can assign a distance between two letters in the alignment by asking the question "how likely is it that the two letters are related through mutation." One way to answer this question is to compare two probabilities: (1) the probability that the letter in question in one sequence would take on that value as a result of a mutation in the letter in the other sequence, and (2) the probability that the letter in question would randomly take on that value. If we assume that each letter in a sequence takes on its value independently of all other letters in the sequence then the likelihood of the sequences, rather than letters being mutated forms of each other can be written in terms of the products of individual probabilities. So given two sequences

$$X = [x_1, x_2, \ldots, x_n]$$

and

$$Y = [y_1, y_2, \ldots, y_n]$$

where we assume that the sequences have been aligned such that the two sequences (with any necessary gaps) are the same length, we can compare the probability that a letter y_i arose due to a mutation of the letter x_i with the probability that the letter y_i occurred at random by computing the ratio

$$\frac{P(y_i|x_i)}{P(y_i)}$$

The numerator is the probability that the ith element of the sequence Y will be y_i given that the ith element of the sequence X is x_i. The denominator is the probability that the ith element of the sequence Y is y_i regardless of what value the sequence X takes on. If this ratio is greater than one that will indicate that the probability of the two sequences being related at that point is higher than the probability that the two nucleotides or amino acids at that location are unrelated. If the ratio is less than one the reverse is the case. Assuming mutations occur independently we can compute the likelihood of a particular alignment of the sequences X and Y to be

$$L(X, Y) = \prod_{i=1}^{n} \frac{P(y_i|x_i)}{P(y_i)}$$

If the two sequences are not related the terms in the product will mostly have a value less than one and the likelihood value $L(X, Y)$ will be small. It is generally more computationally convenient to evaluate the log-likelihood because the log-likelihood does not require products of increasingly small numbers that would result in an underflow in most computers. The logarithm of the likelihood can be computed as a sum:

$$\log L(X, Y) = \sum_{i=1}^{n} \log \left[\frac{P(y_i|x_i)}{P(y_i)} \right]$$

Now a term in the summation is positive if the ratio $P(y_i|x_i)/P(y_i)$ is greater than one and negative otherwise. We can also expand the logarithm to write the log-likelihood function as

$$\log L(X, Y) = \sum_{i=1}^{n} [\log P(y_i|x_i) - \log P(y_i)]$$

Another way to derive the likelihood function is to compare the joint probability of x_i and y_i taking on particular values assuming they occur independently and if their occurrence is not independent. In the latter case, we can write the joint probability as

$$P(x_i, y_i) = P(y_i|x_i)P(x_i)$$

and in the former case, we can write the joint probability as a product of the marginal probabilities.

$$P(x_i, y_i) = P(x_i)P(y_i)$$

Taking the ratio of these two

$$\frac{P(y_i|x_i)P(x_i)}{P(y_i)P(x_i)}$$

$P(x_i)$ cancels out and we are left with the same expression as before. This approach renders the term in the log likelihood function as the log-odds ratio that the letters are dependent.

In any case, in order to compute the log-likelihood we need the probabilities that go into making up the expression. For protein sequences, Margaret Dayhoff, one of the pioneers of bioinformatics, developed the first set of values to be used in log-likelihood calculations in 1978 [1]. She and her coworkers gathered amino acid sequences that differed in 1% of their residues, aligned them, and then computed the various probabilities. The reason for picking sequences that were so similar was that the differences could reasonably be assumed to be due to mutations. The result was a matrix called the PAM1 matrix, where PAM stands for Percent Accepted Mutation, and 1 denotes the fact that the sequences from which these probabilities were obtained differed by 1%. By assuming that repeated mutations would follow the same rules we can obtain the PAMk matrices for different values of k. Thus the PAM2 matrix is simply the PAM1 matrix multiplied by itself, and the PAMk matrix is the PAM1 matrix raised to the kth power.

The initial approach of Dayhoff and colleagues was a very sensible one. Sequences that are very similar to each other can be easily aligned and we can get very consistent estimates for the various probabilities. However, the assumption that likelihood ratios for scenarios where there was more disagreement between the sequences could be found simply by taking repeated products of the PAM1 matrix was not borne out. This problem was addressed by Henikoff and Henikoff [2]. They collected divergent protein sequences and then estimated the probabilities based on blocks of residues that were conserved. The matrices derived

Table 4.1 The BLOSUM62 Matrix

	A	R	N	D	C	Q	E	G	H	I	L	K	M	F	P	S	T	W	Y	V
A	4																			
R	−1	5																		
N	−2	0	6																	
D	−2	−2	1	6																
C	0	−3	−3	−3	9															
Q	−1	1	0	0	−3	5														
E	−1	0	0	2	−4	2	5													
G	0	−2	0	−1	−3	−2	−2	6												
H	−2	0	1	−1	−3	0	0	−2	8											
I	−1	−3	−3	−3	−1	−3	−3	−4	−3	4										
L	−1	−2	−3	−4	−1	−2	−3	−4	−3	2	4									
K	−1	2	0	−1	−3	1	1	−2	−1	−3	−2	5								
M	−1	−1	−2	−3	−1	0	−2	−3	−2	1	2	−1	5							
F	−2	−3	−3	−3	−2	−3	−3	−3	−1	0	0	−3	0	6						
P	−1	−2	−2	−1	−3	−1	−1	−2	−2	−3	−3	−1	−2	−4	7					
S	1	−1	1	0	−1	0	0	0	−1	−2	−2	0	−1	−2	−1	4				
T	0	−1	0	−1	−1	−1	−1	−2	−2	−1	−1	−1	−1	−2	−1	1	5			
W	−3	−3	−4	−4	−2	−2	−3	−2	−2	−3	−2	−3	−1	1	−4	−3	−2	11		
Y	−2	−2	−2	−3	−2	−1	−2	−3	2	−1	−1	−2	−1	3	−3	−2	−2	2	7	
V	0	−3	−3	−3	−1	−2	−2	−3	−3	3	1	−2	1	−1	−2	−2	0	−3	−1	4

by Henikoff and Henikoff are called BLOSUMxx matrices. BLOSUM stands for Block Substitution Matrix and xx denotes the percent of the sequence that was conserved. So the BLOSUM62 matrix shown in Table 4.1 was obtained from sequences in which 62% of the residues were conserved. According to the BLOSUM62 matrix we can align an Alanine with an Alanine for a score of 4 but an Alanine aligned to a Phenylalanine will give us a negative score of -2.

4.4 Alignment Approaches

Given a scoring mechanism. we can now talk about the different ways in which a scoring mechanism can be used to perform an alignment. There are two main types of approaches for aligning sequences. One type uses dynamic programming, which provides the optimal solution, but for applications like database searches, these approaches can be time-consuming. The second type of approaches are word-based approaches which while not optimal are significantly faster and are popular for searches.

4.4.1 Dynamic Programming Approaches

Dynamic programming provides a recursive solution when a complex problem can be divided into smaller similar subproblems. The additive nature of scoring alignments renders the sequence alignment problem suitable for a dynamic programming approach where the same solution is recursively applied to subproblems to solve the main, complex problem. The "additive" property comes from the fact that the score of an alignment is the sum of the scores of the individual columns of the alignment.

We can define the alignment scores as a function $S(x, y)$ where x and y are either letters from the alphabet of the sequences or the gap symbol '-.' We exclude the pair (-,-) as no column of an alignment can consist of gaps only. The range of S is typically the real numbers. We define $S(x, y) = S(y, x)$ as the score of aligning the symbol x with the symbol y, $S(-, x) = S(x, -)$ is defined as the score of aligning symbol x with a gap. Note that the scoring function is independent of factors such as the position of the column or the order of the sequences. A slight improvement to gap scoring is known as the 'affine gap scoring' where different scores are assigned to 'opening gaps' which is the first gap that we introduce into a sequence and 'extending' gaps which extend an already introduced gap.

The dynamic programming framework for sequence alignment works because of the following observation. Let's assume we have two sequences p and q of length n_1 and n_2, respectively. Let $p(i)$ denote the ith letter of the sequence p and let $p(i..j)$ denote the subsequence of p between the positions i and j (inclusive). The optimum alignment between the sequences p and q must have one of the following as its last column: $p(n_1)$ is aligned with $q(n_2)$, $p(n_1)$ is aligned with a gap, or a gap is aligned with $q(n_2)$. Let $o[a, b]$ denote the score of the optimum global alignment between two (sub)sequences a and b. Let us now investigate the score of the optimum global alignment between p and q in all three cases:

Case 1 ($p(n_1)$ is aligned with $q(n_2)$):

$$o[p, q] = o[p(1..n_1 - 1), q(1..n_2 - 1)] + S(p(n_1), q(n_2))$$

Case 2 ($p(n_1)$ is aligned with a gap):

$$o[p, q] = o[p(1..n_1 - 1), q(1..n_2)] + S(p(n_1), -)$$

Case 3 (a gap is aligned with $q(n_2)$):

$$o[p, q] = o[p(1..n_1), q(1..n_2 - 1)] + S(-, q(n_2))$$

Since $o[p, q]$ is the maximum possible attainable score when we align the sequences p and q, it follows that

$$o[p, q] = \max \{o[p(1..n_1 - 1), q(1..n_2 - 1)] + S(p(n_1), q(n_2)), o[p(1..n_1 - 1), q(1..n_2)] ,$$
$$+ S(p(n_1), -)o[p(1..n_1), q(1..n_2 - 1)] + S(-, q(n_2))\}$$

It follows that we can extend this idea to any two prefixes of p and q. In other words, we can say:

$$o[p(1..i), q(1..j)] = \max \{o[p(1..i - 1), q(1..j - 1)] + S(p(i), q(j)),$$
$$o[p(1..i - 1), q(1..j)] + S(p(i), -), o[p(1..i), q(1..j - 1)] + S(-, q(j))\}$$

The approaches for finding the optimum global alignment between two sequences using dynamic programming are based on this recursive formula.

The most popular dynamic programming approach was introduced by Saul Needleman and Christian Wunsch in 1970 [3] for aligning protein sequences. The Needleman–Wunsch technique is a *global alignment* algorithm which means that the entirety of both sequences is aligned to each other with the inclusion of spaces or gaps as necessary. Later we will look at local alignments where the focus is on finding the best matching *regions* of the sequences.

The Needleman–Wunsch Algorithm

The Needleman–Wunsch algorithm as initially introduced was quite cumbersome and various refinements have been introduced. The algorithm as currently used can be divided into three steps, initialization, forward progression, and traceback. The easiest way to describe the algorithm is to use an example. Let's align the sequences GAGACAT and GATCA using the scoring matrix S shown below and a gap penalty of -1. A gap penalty means that when

	A	C	G	T
A	2	0	1	0
C	0	2	0	1
G	1	0	2	0
T	0	1	0	2

we align a letter with a gap we incur a penalty of -1. Also, the alignment of two letters or bases bears different scores depending on the letters. Aligning two bases that are the same gives us a score of 2. Aligning a purine with a different purine, or a pyrimidine with a different pyrimidine garners a score of 1. Aligning a purine with a pyrimidine results in a score of 0, and as we mentioned before inserting a gap garners a negative score.

Let's define the matrix W as

$$W[i, j] = o\,[p(i..i - 1), q(1..j - 1)]$$

We begin with a matrix W of size $n + 1 \times m + 1$ where the sequences are of size n and m.

		G	A	G	A	C	A	T
	0							
G								
A								
T								
C								
A								

Now we fill in the matrix proceeding away from the top left-hand corner using the algorithm

$$W[i, j] = max\left\{W[i - 1, j - 1] + S(x_i, y_j), W[i - 1, j] + \delta, W[i, j - 1] + \delta\right\}$$

where $S(x_i, y_i)$ is the score when aligning the letter x_i with the letter y_i and δ is the gap penalty.

The first column and the first row correspond to inserting gaps at the beginning of the first and second sequence respectively. Therefore, the score for each is given by the gap score added to the accumulated score.

		G	A	G	A	C	A	T
	0	← −1	← −2	← −3	← −4	← −5	← −6	← −7
G	↑ −1							
A	↑ −2							
T	↑ −3							
C	↑ −4							
A	↑ −5							

Now begin to fill in the rest of the matrix. The corner cell corresponds to aligning G with G. As the diagonal term has a score of zero the score in this cell will be 2.

		G	A	G	A	C	A	T
	0	← −1	← −2	← −3	← −4	← −5	← −6	← −7
G	↑ −1	↖ 2						
A	↑ −2							
T	↑ −3							
C	↑ −4							
A	↑ −5							

Along with filling in the accumulated score we also note which of the three options, namely gap in the top sequence, gap in the bottom sequence or extension of alignment, provided the maximum score. In this case, a gap in the top or bottom sequence would have resulted in a score of -1 while aligning the first G in GAGACAT and GATCA resulted in a score of $+2$.

Now complete the next larger square:

		G	A	G	A	C	A	T
	0	← −1	← −2	← −3	← −4	← −5	← −6	← −7
G	↑ −1	↖ 2	← 1					
A	↑ −2	↑ 1	↖ 4					
T	↑ −3							
C	↑ −4							
A	↑ −5							

Continuing in this manner we can complete the matrix as follows:

		G	A	G	A	C	A	T
	0	← −1	← −2	← −3	← −4	← −5	← −6	← −7
G	↑ −1	↖ 2	← 1	← 0	← −1	← −2	← −3	← −4
A	↑ −2	↑ 1	↖ 4	← 3	↖ 2	← 3	← 2	← 1
T	↑ −3	↑ 0	↑ 3	↖ 4	← 3	↖ 3	↖ 3	↖ 4
C	↑ −4	↑ −1	↑ 2	↖ 3	↖ 4	↖ 5	← 4	← 3
A	↑ −5	↑ −2	↑ 1	↖ 3	↖ 5	↖ 4	↖ 7	← 6

In order to use this matrix to perform the alignment we begin at the bottom right corner and follow the arrows back to the top left corner.

		G	A	G	A	C	A	T
	0	←−1	←−2	←−3	←−4	←−5	←−6	←−7
G	↑−1	↖2	←1	←0	←−1	←−2	←−3	←−4
A	↑−2	↑1	↖4	←3	↖2	←3	←2	←1
T	↑−3	↑0	↑3	↖4	←3	↖3	↖3	↖4
C	↑−4	↑−1	↑2	↖3	↖4	↖5	←4	←3
A	↑−5	↑−2	↑1	↖3	↖5	↖4	↖7	←6

The alignment is therefore:

$$\text{G A G A C A T}$$
$$\text{G A T} - \text{C A} -$$

In order to implement this algorithm we need the matrix of scores and the matrix of back−pointers. We can do this by using a three-dimensional array $W[i, j, k]$ where i and j range over the lengths of the two sequences and k takes on only two values. For example, $W[i, j, 0]$ could be the score at location (i, j) and $W[i, j, 1]$ contains the pointer. There are only three possible places for the pointer to point to, up, left, and diagonal. We can encode these with 0,1, and 2. For the example shown above $W[i, j, 1]$ would be:

		G	A	G	A	C	A	T
0	1	1	1	1	1	1	1
G	0		1	1	1	1	1	1
A	0	0	2	1	2	1	1	1
T	0	0	0	2	1	2	2	2
C	0	0	0	2	2	2	1	1
A	0	0	0	2	2	2	2	1

Using such a matrix we could generate the final alignment using the following pseudocode: Assume $n \geq m$

```
Begin with i ← n, j ← m
While( i ≥ 0 and j ≥ 0)
{
    pointer = W[i,j,1]
    if pointer = 2,
        align1[i] = sequence1[i], align2[i] = sequence2[j]
        i ← i − 1, j ← j − 1
    if pointer = 1,
        align1[i] = sequence1[i], align2[i] = -
        j ← j − 1
    if pointer = 0
        align1[i] = -, align2[i] = sequence2[j]
        i ← i − 1.
}
```

Local Alignment—Smith–Waterman Algorithm

The Needleman–Wunsch algorithm is a global alignment algorithm, that is it aligns the entire sequences. This can mean that subsequences of the sequences being aligned which are well matched to each other may not get aligned because their alignment may have a negative effect on the overall alignment. This might be what we want. On the other hand, there might be situations where we are interested in finding those portions of the sequences that match each other the best. The approach that handles this case is known as "local alignment." In Fig. 4.6, we show both the global and local alignment of two sequences. The "motif" that is identical in the two sequences is missed by the global alignment approach. The local alignment disregards the scores of the aligned portions other than the motif and thus can successfully identify it. The Smith–Waterman algorithm [4], which can be obtained using only slight modifications of the Needleman–Wunsch algorithm provides such an alignment. The modifications are the inclusion of a row and a column at the top and left of the alignment matrix which are filled with zeros, a requirement that no traceback pointer be recorded when the score of a particular matrix element is zero and the update rule is modified to include one more term as follows:

$$W[i, j] = max \left\{ W[i - 1, j - 1] + S(x_i, y_j), W[i - 1, j] + \delta, W[i, j - 1] + \delta, 0 \right\}$$

Let's use the Smith–Waterman algorithm to align GAGACAT and GATCA using the same scoring matrix as before. We begin with the matrix of size $n \times m$

		G	A	G	A	C	A	T
	0							
G								
A								
T								
C								
A								

and fill in the matrix proceeding from the top left-hand corner using the Smith–Waterman algorithm.

The first column and the first row correspond to inserting gaps at the beginning of the first and second sequence respectively. In the case of the Needleman-Wunsch algorithm the score for each was given by the gap score added on to the accumulated score, however, as those scores would be negative and 0 is greater than any negative number the scores become 0

Now begin to fill in the rest of the matrix. The corner cell corresponds to aligning G with G. As the diagonal term has a score of zero the score in this cell will be 2. For the next cell over to the right the score for aligning A with G is 1. As the diagonal term is zero $W[i - 1, j - 1] + S(x_i, y_j)$ will be 1 (this would mean we had aligned the G preceding the A with a gap). If we aligned A with a gap the gap penalty would result in a score of 1. If

```
g A T C t A A t G c a g c G G A t C C a G c c A - A c G - - g g - - A c c - A a G
| - - | - | - | - | - - - | - - - - | - - | | - - - - | - | -
c A T C - A A c G - - - - G G A - C C - - G t g A t A - G t c t t t c t A a a t A a G
```

```
g a t c t a a t g c a g c G G A t C C a G c c A - A c G - - g g - - A c c - A a G
```

```
g a t c t a a t g c a g c c C A A C G G G A C C a a g
| | | | | | | | | | | |
c a t c A A C G G G A C C g t g a t a g t c t t t c t a c a t a g g
```

Fig. 4.6 Global and local alignment of two sequences

		G	A	G	A	C	A	T
	0	0	0	0	0	0	0	0
G	0							
A	0							
T	0							
C	0							
A	0							

we inserted a gap before A we would get a score of -1. The maximum of these is 1 so the score is 1. Continuing in a like fashion we can fill in the matrix.

		G	A	G	A	C	A	T
	0	0	0	0	0	0	0	0
G	0	↖2	←1	↖2	←1	0	↖1	0
A	0	↖1	↖4	←3	↖4	←3	↖2	↖1
T	0	0	↖3	↖4	←3	↖5	←4	↖4
C	0	0	0	↖3	↖4	↖5	↖5	←4
A	0	↖1	↖2	↑2	↖5	↖4	↖7	←6

To find the best local alignment we find the maximum value in the matrix and trace back until we hit a 0. The maximum value being 7 the best local alignment is

$$G\ A\ G\ A\ C\ A$$
$$G\ A\ -\ T\ C\ A$$

We can see the difference between a global and local alignment in Fig. 4.6 using somewhat longer sequences:

p TCCCAGTTATGTCAGGGGACACGAGCATGCAGAGAC
q : TCCCAGTTATGTCAGGGGACACGAGCATGCAGAGACAATTGCCGCCGTCGTTTTCAGCAGTTATGTCAGATC

Semi-global Alignment

The global alignment is used when we want an end-to-end alignment of the sequences p and q. The local alignment is used when we want to find the best match between substrings in p and q. Another situation is when we want to find overlaps between sequences such as when we are putting together sequences generated during shotgun sequencing. We can get such an alignment if we do not penalize for gaps at the beginning or end of p and q. We can accomplish this as we did in the case of local alignments by initializing the first column and the row of the W matrix with zeros. Because we do not need end-to-end alignment we begin the traceback at the maximum value of W on the last column or the last row of the scoring matrix. The various options are summarized in Fig. 4.7.

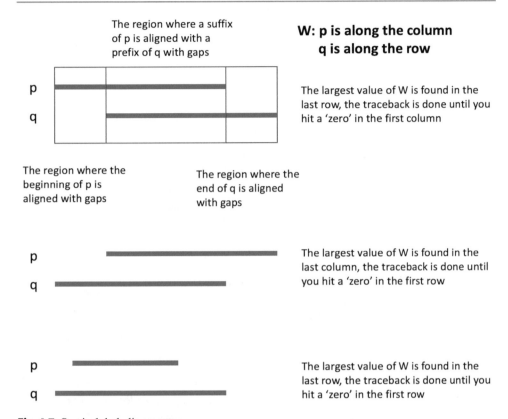

Fig. 4.7 Semi-global alignments

4.5 Word Based Approaches

The Needleman–Wunsch and the Smith–Waterman algorithms both require computations on the order of the product of the length of the sequences. For reasonably long sequences and for a database of reasonable size against which to compare the query sequence this can result in a significant amount of computation. To get around this computational bottleneck Lippman and colleagues introduced two algorithms FASTP and FASTA. FASTP was the first to be introduced and was written for protein sequences—hence the P. FASTA was a generalization of FASTP and could be used for both protein and nucleotide sequences—the A stands for All.

4.5.1 FASTP and FASTA

When we are comparing two sequences we can speed up the comparison by looking at groups or subsequences of letters in the sequence instead of comparing the sequences letter by letter. The first step in such a comparison is to find the common subsequences. Given two sequences S_1 and S_2 we can find all k-tuples of length $ktup$ in the sequences by examining the sequence through a sliding window of length $ktup$ and assigning a numerical value to each subsequence of length $ktup$. There several ways to do this. Here is one for an oligonucleotide. We first convert each letter to a number, for example, $A \to 0$, $G \to 1$, $C \to 2$, and $T \to 3$. Then for each sequence within a window of size $ktup$ we treat the numbers as the digits of a number base 4 which we use as an index. Consider the sequence $GAGACAT$ and let's pick $ktup$ to be 3. The sequence of bases will be translated to the sequence of numbers 1010203. Now put the window around the first 3 numbers $\boxed{101}$0203. Treating 101 as the three digits of a base 4 number we can translate this to a decimal index as $1 \times 4^2 + 0 \times 4^1 + 1 \times 4^0 = 1 \times 16 + 0 \times 4 + 1 \times 1 = 17$. So, the number 17 represents the sequence GAG. Moving the box by one base we get $1\boxed{010}203$ and this subsequence is translated to the number 4. Continuing in this fashion we get

Window	Index	Corresponding sequence
$\boxed{101}$0203	17	GAG
$1\boxed{010}203$	4	AGA
$10\boxed{102}03$	18	GAC
$101\boxed{020}3$	8	ACA
$1010\boxed{203}$	35	CAT

With $ktup = 3$ we can get only 64 different indices. We can construct a table of size 64 and note the presence or absence of a particular subsequence and its location. For example, if we wanted to know whether GAC is contained in the sequence we can look up the index 18 in the table. In general, for FASTA the value of $ktup$ is 2 or 3 for amino acids. Once we have a table built for one sequence we can compare another query sequence to it simply by computing the indices and checking in the table to see if a correspondence exists. Once correspondences have been identified they can be connected if the intervening interval is small. For the sequences in a database we can construct a table for each sequence ahead of time. When we get a query sequence we can compute the table for it and compare with the tables for the database.

The original FASTA algorithm consists of four phases. We will describe these phases using a toy example comparing the sequences GAGACATAG and TGAGACAAT. For our toy example, we will use $ktup = 2$. In Fig. 4.8 we show the dotplot comparison of these two sequences.

Let's assume GAGATAGAG is the reference sequence in the database. Converting the letters to numbers and using a sliding window of two we get the following table of tuples in the sequence

This being a very short sequence not all the entries in the table are filled. Now let's take the query sequence and generate the numerical hashes for each pair of letters. The first pair is TG which corresponds to a hash value of $3 \times 4 + 1 = 13$. The location 13 in our table is empty so there is no match in the database sequence. The next pair starting at the second location is GA with a hash value of 4. There are two matches to this in the table at locations

Fig. 4.8 Dot plot of sequences GAGACATAG and TGAGACAAT

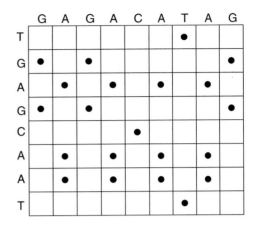

Sequence	Hash	Locations
AA	0	
AG	1	2
AC	2	4, 8
AT	3	6
GA	4	1, 3
⋮		
CA	8	5
⋮		
TA	12	7
⋮		

Fig. 4.9 The hotspots for database sequence GAGACATAG and query sequence TGAGACAAT

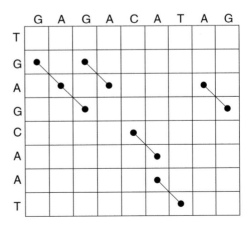

1 and 3. We can write these matches as pairs (2, 1) and (2, 3) where the first element in the pair is the location in the query sequence and the second element is the location of the matching ktuple in the database sequence. These matches are called *hotspots*. For this particular query and database sequences the hotspots are (2, 1), (2, 3), (3, 2), (5, 5), (7, 6). In Fig. 4.9 we show a pictorial representation of the hotspots. In the next step, we want to find hotspots that can be combined without using insertions or deletions. This is the same as finding diagonals in the dot plot matrix which have a significant number of hotspots. An easy way to do this is to notice that the coordinates of the hotspot also tell us if they are on the same diagonal. Given a location (i, j) the diagonal can be indexed by $i - j$. The main diagonal then corresponds to an index of 0. Diagonals above it have negative indices and diagonals below it have positive indices. We can find the diagonals for the hotspots in our toy example as

Sequence	Hotspot	Diagonal Index
GA	(2, 1)	1
GA	(2, 3)	−1
AG	(3, 2)	1
CA	(5, 5)	0
AT	(7, 6)	1

If the distance between hotspots is less than a threshold they are combined into a single region. Otherwise, the first region is saved and we begin a new region on the same diagonal. Looking at our toy example we can see we have three hotspots on the diagonal with index 1 (the diagonal just below the main diagonal). If we took our alignment according to this diagonal we would get the alignment

$$G\ A\ G\ A\ C\ A\ T$$
$$G\ A\ G\ C\ A\ A\ T$$

Notice that there are no insertions or deletions in this alignment. Each of these diagonal regions is initially scored using a positive score for hotspots and negative scores for mismatches. In the FASTA algorithm, the top ten scoring local regions are extracted. These regions are rescored using a scoring matrix such as BLOSUM50 for proteins or a scoring matrix similar to the one we used earlier for nucleotides. The score of the best scoring local region is referred to as *init1*. FASTA then checks to see if the best scoring regions in different diagonals can be joined together with gaps. The score of these joined regions is the sum of the scores of the initial local regions plus gap penalties. In our toy example if we had not joined the hotspots on the diagonal with index 1 we would get the alignment shown pictorially in Fig. 4.10. This corresponds to the alignment

$$G\ A\ G\ A\ C\ -\ A\ T$$
$$G\ A\ G\ -\ C\ A\ A\ T$$

Fig. 4.10 The alignment with gaps for database sequence GAGACATAG and query sequence TGAGACAAT

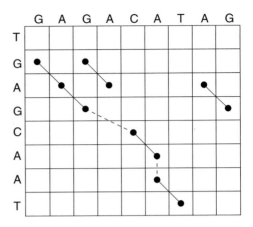

Finally, the FASTA algorithm uses a modified Smith–Waterman algorithm to find the best alignment. In this modified form of the algorithm instead of filling in the entire dynamic programming matrix, only a band of locations along the best alignment so far is evaluated.

4.6 BLAST

BLAST (Basic Local Alignment Search Tool) is probably the most commonly used bioinformatic tool. Since its introduction in 1990, it has spawned a number of different variants. With its many variants, it would probably be more accurate to refer to BLAST as a set of tools. While the general algorithm used is very similar when used for protein sequences or nucleotide sequences it is easier to just look at the algorithms for the two types of sequences separately. We will begin with *blastn* which is used for nucleotide sequences and then look at *blastp* which is used for protein sequences. We will then briefly mention the most popular variants.

4.6.1 Blastn

When dealing with nucleotide sequences BLAST uses a window size or word-length W of 11. Just as in FASTA, we find the hotspots for a query sequence q in the database. In order to do this efficiently the database has to have the associated lookup tables already computed. In the vernacular of BLAST, these exact matches are called seeds. These seeds are then extended in both directions allowing gaps and substitutions. At each step, the score $\sigma(s, t)$ is computed using a scoring matrix, where s is the nucleotide fragment from the query sequence and t is the segment from the database. This extension continues as long as the score $\sigma(s, t)$ is greater than a specified threshold T. Once the score falls below the threshold the extension is stopped and the sequence is trimmed back to where the score was maximum to obtain maximal segment pairs (MSPs). Consider the case where the word-length W is 3, the threshold is 4, the query sequence is GAGACATAGAGT, and the database sequence is GAGACTTACTCG. Let's assign a score of one to a match and -1 to a mismatch. We can then score the extensions as:

$$
\begin{array}{ll}
s: & \text{G A G A C A T A G A G T} \\
t: & \text{G A G A C T T A C T C G} \\
\sigma(s,t): & 1\ 2\ 3\ 4\ 5\ 4\ 5\ 6\ 5\ 4\ 3\ 2
\end{array}
$$

We can see that the score of the match drops below 4 at the last C so we stop extending the seed when we get to GAGACTTCT. At this point, $\sigma(s,t)$ is 4. If we now trim off the T and C from the end we get a score of 6 and an MSP of (GAGACATA,GAGACTTA). Once all MSPs have been found the highest scoring pairs with score greater than some threshold S are identified as HSPs. The statistical significance of these High-Scoring Segment Pairs (HSPs) are computed and the significant HSPs are examined to see if they can be combined. The statistical significance in BLAST is reported in terms of the Expect value or E-value. This is the number of hits with this score one expects to see at random in this particular database. So if you have an E-value of 1 then you would expect to see a one hit with this score in this database just by chance. If the E-value is 0.1 then you would expect to see a hit with this score just by chance once in ten such searches. If the E-value is 10^{-50} it is highly unlikely that this was a random match and this particular match has high significance. The closer to zero the E-value is the higher the significance.

4.6.2   Blastp

For protein sequences the word length W is generally much smaller; $W = 3$ is a commonly used word length. Once the words from the query sequence have been obtained, for each word a list of words that have a similarity score greater than some specified threshold is compiled. Let's suppose our query sequence was ARE and the similarity threshold was 10. Using the BLOSUM62 matrix shown in Table 4.1 the following sequences have a distance of 10 or greater to ARE: ARD, ARQ, ARK, AQE, AKE, CRE, GRE, SRE, TRE, and VRE. Rather than just using ARE to find the seeds in the database, all of these sequences are used to generate the seeds. The rest of the process is algorithmically the same as in blastn.

The score of an alignment depends on the distance between components of the sequence. The simplest distance is the Hamming distance in which if two letters in a particular position of an alignment are the same the Hamming distance at that position is 0 and if they are different the Hamming distance at that position is 1. The distance between the sequences is the sum of the distances of the positions and we would prefer the alignment with a smaller Hamming distance. A generalization of this would be the *edit distance*.

4.7 Edit Distance

The different algorithms we have looked at maximize the score of an alignment. Another way of comparing pairwise alignments is to look at the cost of converting one sequence into the other based on the alignment. This cost is called the edit distance. In this case, the best alignment is one that has the minimum edit distance or Levenshtein distance after the Soviet mathematician Vladmir Levenshtein who proposed it in the context of error correcting codes [5].

The minimum edit distance or the *Levenshtein distance* is the smallest number of single letter changes you need to make in one sequence to obtain the other sequence. Where the sequences being compared are the same length this would simply be the number of places where the sequences are different. Where the sequences are of different lengths the calculation can become slightly more involved. Suppose we are finding the edit distance between two sequences; a sequence S of length N and a sequence R of length M. Let's define a distance matrix $\mathbf{D}$ of size $N \times M$. The location (i, j) of this matrix denotes the edit distance between the first i elements of S and the first j elements of R. The first row and the first column of the matrix correspond to the empty string. Because of this we will index our rows and columns with 0. So $D[0, j]$ corresponds to the edit distance when the first j elements of the R are aligned to the empty string. Clearly, this value will be j. The N, Mth element of this matrix will contain the edit distance between the two sequences. At each step in our alignment we can do one of three things. We can take the alignment up to the current point and simply align the next element in each sequence, we can insert a gap into the sequence S; we can insert a gap into the sequence R. For the first case where we take the alignment up to the current point and simply align the next element in each sequence, if these two elements are the same then the additional cost is going to be zero; if they are different the additional cost will be 1. In this case, the distance matrix D will be updated as

$$D[i, j] = D[i - 1, j - 1] + HD(S[i], R[j])$$

where $HD(a, b)$ is the Hamming distance between a and b which is 0 if $a = b$ and 1 otherwise. If we insert a gap into the sequence S this will mean an additional cost of 1 and the distance matrix will be updated as

$$D[i, j] = D[i, j - 1] + 1$$

If we insert a gap into the sequence R again this will mean an additional cost of 1 and the distance matrix will be updated as

$$D[i, j] = D[i - 1, j] + 1$$

Because we are computing the minimum edit distance we select the action corresponding to the minimum of these three and the distance matrix update becomes

$$D[ij] = min(D[i-1, j-1] + HD(S[i], R[j]), D[i-1, j]+1, D[i, j-1]+1)$$

To see how this works let's calculate the edit distance between BACKYARDS and BAR-RAYAR. We let the first sequence, in this case BACKYARDS, define the rows, and the second sequence BARRAYAR define the columns.

		B	A	R	R	A	Y	A	R
	0	1	2	3	4	5	6	7	8
B	1								
A	2								
C	3								
K	4								
Y	5								
A	6								
R	7								
D	8								
S	9								

The element $D[0, 1]$ is the edit distance between an empty string and the first letter of the BARRAYAR sequence. To go from B to an empty string requires one deletion so the edit distance is 1. The element $D[0, 2]$ is the minimum number of edits required to go from the two letter prefix of BARRAYAR, BA and the empty string, which is 2. And so on. In the same way the entry $D[j, 0]$ is the number of edits required to go from the j letter prefix of BACKYARDS and the empty string. The entry $D[1, 1]$ corresponds to the alignment of B of BARRAYAR and B of BACKYARDS so $D[1, 1] = 0$. The entry $D[1, 2]$ corresponds to the alignment of BA from BARRAYAR and B from BACKYARDS. We can do this in three different ways. We can insert a gap after BA to align with B. This would give us the alignment

B A

B

and an edit distance of 3. We could insert a gap after B which would give us the alignment

B A

B

and an edit distance of 1. Or, we could align A to B to give us the alignment

B A

B

and an edit distance of 2. The minimum of 3, 1, and 2 being 1, $D[1, 2] = 1$. Rather than go through each possibility in detail we could simply have used the update equation

$$D[i, j] = min \left(D[i - 1, j - 1] + HD(S[i], R[j]), D[i - 1, j] + 1, D[i, j - 1] + 1\right)$$
$$= min \left(1 + 1, 2 + 1, 0 + 1\right)$$
$$= 1$$

We fill the remaining entries recursively using the equation above to get

		B	A	R	R	A	Y	A	R
	0	1	2	3	4	5	6	7	8
B	1	0	1	2	3	4	5	6	7
A	2	1	0	1	2	3	4	5	6
C	3	2	1	1	2	3	4	5	6
K	4	3	2	2	2	3	4	5	6
Y	5	4	3	3	3	3	3	4	5
A	6	5	4	4	4	3	4	3	4
R	7	6	5	4	4	4	4	4	3
D	8	7	6	5	5	5	5	5	4
S	9	8	7	6	6	6	6	6	5

The entry $D[9, 8] = 5$ so the minimum number of changes to convert one sequence into the other would be 5. We can find the alignment by starting at the lower right corner of the matrix and tracing back to the (1,1) element by always going to the smallest entry among the element directly above, directly to the left, and at the left diagonal. If we do this in this matrix we find three possible paths corresponding to the alignments:

$$\begin{array}{l} B\ A\ R\ R\ A\ Y\ A\ R \\ B\ A\ C\ K\ \ \ \ \ Y\ A\ R\ D\ S \end{array}$$

$$\begin{array}{l} B\ A\ R\ R\ A\ Y\ A\ R \\ B\ A\ C\ \ \ \ K\ Y\ A\ R\ D\ S \end{array}$$

$$\begin{array}{l} B\ A\ R\ R\ A\ Y\ A\ R \\ B\ A\ \ \ \ \ C\ K\ Y\ A\ R\ D\ S \end{array}$$

If we do the same with the sequences ARRAKIS and BARRAYAR we get the distance matrix

Looking at the lower right corner of the table we can see that the minimum edit distance between these two sequences is 4. We can find the alignment itself by starting at the lower right corner and following the lowest cost path through the matrix. If we do that we find the alignment to be

$$\begin{array}{l} B\ A\ R\ R\ A\ Y\ A\ R \\ A\ R\ R\ A\ K\ I\ S \end{array}$$

All of these algorithms are for aligning two sequences. In many situations, we are interested in aligning more than two sequences. The problem becomes a bit more complex when we try to do that. We look at the problem of multiple sequence alignment in the next chapter.

		B	A	R	R	A	Y	A	R
	0	1	2	3	4	5	6	7	8
A	1	1	1	2	3	4	5	6	7
R	2	2	2	1	2	3	4	5	6
R	3	3	3	2	1	2	3	4	5
A	4	4	3	3	2	1	2	3	4
K	5	5	5	4	3	2	2	3	4
I	6	6	6	5	4	3	3	3	4
S	7	7	7	6	5	4	4	4	4

4.8   Exercises

1. Assume, we are given the following two sequences:
   S1: CCAGGACTCGATCG S2: CGATCCGATGCG
   Consider the following three possible global pairwise alignments:

   C C A − G G − A C T − C G A − T C G
   C G A T − C C − − G A − − T G − C G

   C C A G G A C T C G A T − C G
   C G A − − T C − C G A T G C G

   C C A G G A C T C G A T C G − − − − − − − − − −
   − − − − − − − − − − − − C G A T C C G A T G C G

   Using the following scoring scheme, determine the best alignment:
   Match: +2 Mismatch: −1 Gap: −2

2. Consider the matrix that keeps scores in the dynamic programming approach to pairwise alignment. Which cell(s) in this matrix would give you the score of the best alignment when you are performing global, local, or semiglobal alignment.

3. What do X and Y mean in PAMX and BLOSUMY scoring matrices?

4. What is the score of aligning Leucine (L) with Serine (S) in the following scoring matrices: BLOSUM45, BLOSUM90, PAM250, PAM30?

5. You have BLOSUM62, BLOSUM80, BLOSUM45, PAM120, PAM60, and PAM250 scoring matrices at your disposal. You also have three groups of sequences: (i) closely related (ii) distantly related (iii) "somewhat" related. When building alignments for sequences in each group, which scoring matrices would you use for each group?

6. What does the E-value of a BLAST output mean?

7. Assume you performed the dot plot of a sequence with itself and obtained the following:
   How would you interpret the three lines shown in this plot?

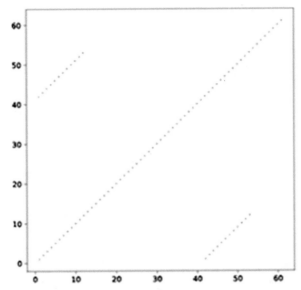

8. Consider the two short red lines in the dot plot figure shown in question 7. Assume they were rotated by 90° pivoted at their midpoints. How would you interpret them now?

9. Take the RefSeq protein ID for the Human VEGFA protein, NP_001020537, and use NCBI blastp to BLAST it against the refseq_protein database restricting the organisms to the house mouse (taxid:10090). Use default parameters. Explore your results. How many hits do you get? What are the E-value and percent identity for your top hit?

10. Repeat question 9, this time using PSI-BLAST and then using DELTA-BLAST. Compare your results from the three approaches.

References

1. M Dayhoff, R Schwartz, and B Orcutt. A model of evolutionary change in proteins. *Atlas of protein sequence and structure*, 5:345–352, 1978.
2. Steven Henikoff and Jorja G Henikoff. Amino acid substitution matrices from protein blocks. *Proceedings of the National Academy of Sciences*, 89(22):10915–10919, 1992.
3. Saul B Needleman and Christian D Wunsch. A general method applicable to the search for similarities in the amino acid sequence of two proteins. *Journal of molecular biology*, 48(3):443–453, 1970.
4. Temple F Smith and Michael S Waterman. Identification of common molecular subsequences. *Journal of molecular biology*, 147(1):195–197, 1981.
5. Vladimir I Levenshtein et al. Binary codes capable of correcting deletions, insertions, and reversals. In *Soviet physics doklady*, volume 10, pages 707–710. Soviet Union, 1966.

Multiple Sequence Alignment

5

In a nutshell ... When we go from aligning two sequences to aligning more than two (or in troll count many [1]) things become more complicated. We examine the complications and the ways around the complications.

5.1 Definition and Use

A multiple sequence alignment (MSA) is an alignment of three or more sequences. Given n strings $S1, ..., Sn$ $(n > 2)$ over an alphabet that does not contain the gap symbol, we insert gaps into the sequences so that their lengths become the same. A construction of these sequences into a matrix form where the rows represent the "gapped" sequences constitutes the MSA of the sequences. The matrix representation enables us to define "columns" of the MSA. By definition, no columns of an MSA can contain gap symbols only. In Fig. 5.1, we show two sample MSAs of three sequences.

MSAs provide us with a richer perspective than what we could achieve with pairwise alignment. Often, we are interested in identifying columns with high conservation. These are the columns where an overwhelming percentage of the symbols are the same. For example, in Fig. 5.1 (MSA2), we highlighted in green the columns where all the symbols are the same; one column consists of Cs only and the other column consists of Gs only. These columns could be interesting from an evolutionary and functional perspective as they represent regions that are "conserved" among the sequences. We look for regions where such conservation is observed in consecutive columns so that we can talk about "conserved regions."

Unlike pairwise alignment, MSAs lead to the identification of "motifs" and "domains" across multiple sequences, which may have functional and evolutionary implications. The premise for this is that the functional and evolutionarily important parts of biological sequences should be mutating at a lower rate than their non-functional or evolutionarily

Sequences	MSA1	MSA2
S1 = CTTGAAGG	C T – T G A A G G –	– C T T G A A G G –
S2 = ACCTAGTT	A C C T – A G T T –	A C C T – A – G T T
S3 = CAGGTTGA	– C A G G – T T G A	– C A G G T T G A –

Fig. 5.1 Two sample MSAs obtained for three sequences. Columns with 2 matches are shaded in *blue* and columns with 3 matches are shaded in *green*

unimportant parts. Hence, such regions exhibit a high similarity among different biological sequences and may be captured by identifying the conserved regions in the MSA of those sequences.

For example, by looking at the MSA of the promoter regions of a set of genes, we may identify motifs that are very similar between these sequences. Such motifs may imply transcription factor binding sites, which in turn may let us infer if these genes are co-regulated. Similarly, through MSA of proteins, we can identify conserved domains that have functional implications such as binding to DNA, other proteins, chemicals, or amino acids. Furthermore, conserved regions across multiple sequences may imply structural similarities such as finding amino acid subsequences that contribute to alpha-helices or beta-sheets. These findings may help with defining the 3D structure of proteins and inferring their functional capabilities.

In Fig. 5.2, we show part of the MSA of the vascular endothelial growth factor A (VEGFA) protein sequences from *Homo sapiens* (human), *Canis lupus familiaris* (dog), *Bos taurus* (cow), *Mus musculus* (mouse), *Rattus norvegicus* (rat), *Gallus gallus* (junglefowl), and *Danio rerio* (zebrafish).

```
Danio    SFaeHSeCqCRmKKD-lpKEieKacRcmapscltsaihtlqpslwtlrrGKtvvpdqggS
Gallus   SFLQHSKCdCRPKKDvKnKQEKKSkRG-------------------KGKGQKRKRKKg
Mus      SFLQHSrCECRPKKD-RtKpEKKSVRG-------------------KGKGQKRKRKKS
Rattus   SFLQHSrCECRPKKD-RtKpEKKSVRG-------------------KGKGQKRKRKKS
Bos      SFLQHnKCECRPKKD-KarQEn------------------------------------
Canis    SFLQHSKCECRPKKD-RarQEKKSIRG-------------------KGKGQKRKRKKS
Homo     SFLQHnKCECRPKKD-RarQEKKSVRG-------------------KGKGQKRKRKKS

Danio    kisqcepcCst CSERRrrLFVQDPeTCqCSCKhseadCrsRQLELNERTCRCDKPRR
Gallus   RYKppSfHCEP CSERRKHLFVQDPQTCKCSCKfTDSRCKsRQLELNERTCRCeKPRR
Mus      RFKSWSVHCEP CSERRKHLFVQDPQTCKCSCKNTDSRCKARQLELNERTCRCDKPRR
Rattus   RFKSWSVHCEP CSERRKHLFVQDPQTCKCSCKNTDSRCKARQLELNERTCRCDKPRR
Bos      -------pCgP CSERRKHLFVQDPQTCKCSCKNTDSRCKARQLELNERTCRCDKPRR
Canis    RYKpWSVpCgP CSERRKHLFVQDPQTCKCSCKNTDSRCKARQLELNERTCRCDKPRR
Homo     RYKSWSVpCgP CSERRKHLFVQDPQTCKCSCKNTDSRCKARQLELNERTCRCDKPRR
```

Fig. 5.2 Part of the MSA of VEGFA protein sequences in seven organisms. The heparin binding domain is highlighted. *Lowercase letters* denote potential insertion/substitution sites; *blue* and *grey* backgrounds denote high and low conservation sites, respectively

In this MSA figure, we highlighted the tail-end of the alignment, which represents the VEGF heparin-binding domain, located at the C-terminus of the protein. Except for the zebrafish, this domain shows high conservation across organisms, unlike other regions of the MSA.

5.2 Scoring

Let's consider an alphabet of size k, $A = \{s1, \ldots, sk\}$, that does not include the gap symbol. We can augment A with the gap symbol and call it $A' = A U \{-\}$. A pairwise alignment scoring function, f, would be defined for $A' x A' - \{-, -\}$ as we would not have two gaps aligned with each other in a pairwise alignment.

In its most general form, this function's domain would have $(k+1)^2 - 1$ entries (couples). On the other extreme, the function would be defined for three types of couples only; we would need match, mismatch, and gap scores. Typically, we assume that the function is symmetrical with respect to its arguments and that aligning the gap with any letter in A would have the same score. Thus, the function f is defined for $(k^2 + k)/2 + 1$ entries representing the lower triangle of the kxk scoring matrix, its diagonal, and the gap penalty. This, for example, is the case when we use a BLOSUM matrix for scoring protein pairwise alignments with a linear gap penalty. As we shall see in the next section, things get more interesting in the case of MSAs.

5.2.1 Sum of Column Scores

A common assumption made in biology is that different positions in a molecular sequence evolve independently of each other. This is the premise for adding the paired (aligned) symbol scores to find the total alignment score in a pairwise alignment. A similar idea applies to the case of MSAs. Consider Fig. 5.3, where we are aligning 4 protein sequences.

There are 7 columns in this MSA and if we assume that the score of this MSA is the sum of its column scores, then we would have

Fig. 5.3 A sample MSA of four protein sequences, $S1, \ldots, S4$

S1:	K	D	–	E	K	S	P
S2:	K	N	V	E	–	F	–
S3:	R	Y	I	–	K	–	P
S4:	K	H	–	L	–	S	P

$$S_{MSA} = \sum_{i=1}^{7} f(C_i)$$

where S_{MSA} is the score of the MSA and $f(C_i)$ is the score of the ith column. In general, if the alphabet size is k and the number of sequences is n, then we would have $n^{(k+1)} - 1$ different possible n-tuples in a column. Hence, a scoring function for a column, f, would need to be defined for $n^{(k+1)} - 1$ entries. This is not practical as for $k = 20, n = 10$, a typical MSA of 10 proteins, we would need a scoring function that has 100 quintillion (billion billion) entries.

A symmetry assumption can bring this number down but still would result in an unreasonably large number. What we mean by the symmetry assumption is that it would make sense to have the function f be symmetrical with respect to its arguments. After all, the order of the sequences in the MSA is somewhat arbitrary. Thus, considering the score of the highlighted 6th column in Fig. 3, we should have the same score if the second and fourth sequences were flipped. In other words, we need to have $f(S, F, -, S)$ to be equal to $f(S, S, -, F)$, or to $f(-, S, F, F)$ for that matter.

But how many unique entries would f need to have even with the symmetry assumption? Let's assume we are aligning n sequences over an alphabet size of k. Then, each column of the alignment is a vector of length n. The entries of this vector are the elements of A'. Due to the symmetry assumption, we only need to count the number of times each symbol occurs in this vector to find the number of different combinations f needs to be defined for.

For example, going with the highlighted column of Fig. 5.3, we can represent that column as "2S,1F,1-". Hence, we would only need to define f for this entry and any reordering of these 4 symbols with the noted counts would be mapped to the same score. This model gives us a way to find the number of entries f needs. Now, think of a $k + 1$ long vector representing the k symbols of A and the gap symbol. We want to insert non-negative integers as this vector's components such that the sum of the components equals n. Each component corresponds to the number of times that symbol shows up in the column of the MSA.

How do we find the number of different ways $k + 1$ non-negative integers add up to n? Consider a sequence of n 1s. We place k commas between these n 1s to define the value of ith integer as the number of 1s between ith and $(i - 1)$st commas.

Assume $n = 7$ and $k + 1 = 5$, simulating the case of aligning 7 DNA sequences. Four cases depicted in Fig. 5.4 show how 4 commas can be placed between 7 1s. For each case, we show the corresponding combination an MSA's column would assume where the augmented alphabet is {A, C, G, T,–}. Note that the number of "positions" in each case is 11 or $n + k$. Hence, the problem can be thought of as finding k positions (where the commas would go) out of $n + k$ total positions. The answer to this is Comb(n+k, k) where Comb represents the combination function. Since a column in an MSA cannot be all gaps, the function that calculates the score of a column in an MSA of n sequences over an alphabet of size k must have Comb(n+k, k)-1 unique entries with the symmetry assumption.

Position:	1	2	3	4	5	6	7	8	9	10	11	A'={A,C,G,T,–}, n = 7, k = 4
Case1	,	,	1	,	1	1	1	1	,	1	1	"0 As, 0 Cs, 1 Gs, 4 Ts, 2 –s"
Case2	1	1	,	1	1	,	1	,	1	,	1	"2 As, 2 Cs, 1 Gs, 1 Ts, 1 –s"
Case3	1	1	1	1	1	,	,	1	,	1	,	"5 As, 0 Cs, 1 Gs, 1 Ts, 0 –s"
Case4	1	,	1	1	1	1	,	1	1	,	,	"1 As, 4 Cs, 2 Gs, 0 Ts, 0 –s"

Fig. 5.4 Four different cases that show how sum of 5 non-negative integers add to 7. Each integer can be thought as the number of symbols in the augmented alphabet A' occur in the column of an MSA. The problem is equivalent to finding k positions in a vector of length n+k

This is still a very large number. For example, considering an MSA of 10 protein sequences $(n = 10, k = 20)$, we would need to define the score of 30,045,014 unique entries a column can assume. Hence, we would need simpler ways to score a column of an MSA.

5.2.2 Defining Column Scores

There are two main approaches to defining the column score of an MSA. One approach uses a metric that takes in the symbols in a column and produces the score. The other approach simply calculates the score of the column as the score of all pairs of symbols in that column.

Metric-Based Approaches

The metrics used to score a column of an MSA try to favor the cases where a low dispersion is seen among the symbols that constitute the column. Some well-known approaches include normalized mean distance score (norMD) [2] and variations on the Hamming-distance [3]. Here, we will describe the entropy approach, which is one of the widely used metrics for this purpose.

We use the term "entropy" in the context of information theory. This metric is sometimes referred to as Shannon's entropy as it was defined by Claude E. Shannon. Entropy measures the amount of information stored in data. It also can be interpreted as the uncertainty in data. Generally, Shannon's entropy converges to the expected Kolmogorov complexity, which is defined as the shortest computer program that outputs the data (Grunweld Vitanyi 2010). The former is viewed as a probabilistic definition of information while the latter is perceived as the algorithmic definition.

Suppose you want to convey the message that a sequence has 1,000,000 As. Instead of typing one million As to explain this data, you can write a simple, short computer program that outputs one million As. Conversely, assume you want to describe (or transmit) a DNA sequence of length one million, but the letters are random. There is no "structure" in this sequence that you can describe by a short code. The only solution would be writing a computer program that outputs the letters one by one.

In the example of all A's, we expect the sequence to have low entropy. We can also view this as the data having a low level of uncertainty or carrying a small amount of information. On the other hand, the random sequence is expected to have high entropy, which could also be interpreted as that data having high uncertainty. Shannon built the probabilistic definition of information, i.e., entropy, by first defining self-information.

Assume we have a space X of k events each occurring with probability of p_i, $i = 1, \ldots, k$. The self-information of an event is defined as

$$SI(i) = log \left(\frac{1}{p_i} \right)$$

The idea is that events that occur with low probability have a high amount of information as SI, the self-information of an event, is inversely proportional to its probability of occurring. This makes sense when you think of some real-life events. For example, if the sun rose from the East tomorrow morning, it would not convey much information to you, as it is almost certainly to happen ($p_i = 1$, $SI = 0$). But if the opposite happened and the sun rose from the West, this event would carry a lot of information for us. When one thinks about events that have low probability of happening, it becomes more reasonable to assume that these events carry more information.

Shannon defined the entropy, H(X), as the expected value of SI:

$$H(X) = E(SI) = \sum_{i=1}^{k} p_i SI = \sum_{i=1}^{k} p_i log(1/p_i) = -\sum_{i=1}^{k} p_i log(p_i)$$

where $0 \log(0)$ is defined as 0. Although the base of the logarithm is arbitrary, it is often chosen to be 2, which results in the unit of entropy to be in *bits*. There is also a constructive approach that lays out the properties of information and shows that the only way $H(X)$ can be defined is as above [4].

This definition of information content in data would explain our sample sequences of length one million in a similar way. Assuming our sequence is a DNA sequence, i.e., there are four "events," the entropy of the sequence where we have one million As would be 0. This is because the only terms that would go in the sum that defines $H(X)$ would be $0\log(0)$s and $1\log(1)$. On the other hand, the random sequence, assuming all 4 letters had the same probability of 1/4, would lead to an entropy of 2 bits ($4 \times [-1/4 \times \log_2(1/4)] = 4 \times [-1/4 \times (-2)] = 2$). This is indeed the highest entropy a sequence of 4 letters can achieve. We can also interpret this as the sequence with the higher entropy possessing more uncertainty.

These definitions and examples show that entropy can be used as a metric to define the score of the column of an MSA. If a column consists of the same symbols, its entropy would be zero. If we diverged from this most certain case by having different symbols in a column,

$$\text{Column entropy score} = - \sum_{i \in \{A,C,G,T\}} p_i log(p_i) \qquad S(MSA) = \sum_{m=1}^{5} S(C_m) = 5.13$$

C	G	T	A	A
C	G	G	A	C
C	G	G	T	G
C	G	G	T	T
C	A	T	C	A

$S(C_1) = 0$

$S(C_2) = - [1/5\log(1/5) + 4/5\log(4/5)] = 0.72$

$S(C_3) = - [2/5\log(2/5) + 3/5\log(3/5)] = 0.97$

$S(C_4) = - \{2\times[2/5\log(2/5)] + 1/5\log(1/5)\} = 1.52$

$S(C_5) = - \{3\times[1/5\log(1/5)] + 2/5\log(2/5)\} = 1.92$

$S(C_m)$: Entropy sore of column m

Fig. 5.5 A sample MSA of DNA sequences and its entropy-based score calculation with a log base of 2. S(MSA): score of the MSA. Low entropy-based scores imply good MSAs

we would monotonically increase the entropy of the column. The score of the MSA would be the sum of the entropies of its columns and the lower the score the better the MSA. A sample MSA and its entropy-based score is shown in Fig. 5.5.

Sum of Pairs

The most popular method used to score the column of an MSA is the sum-of-pairs approach where the column score is the sum of scores of all pairs of symbols that constitute the column. An example is given in Fig. 5.6. The exact same MSA score can also be achieved by the sum of the scores of the induced pairwise alignments. For the sample MSA in Fig. 5.6,

$$\text{Column } i \text{ score} = \sum_{a=1}^{n-1} \sum_{b=a+1}^{n} f(x_i^a, x_i^b)$$

n : number of sequences
f : scoring function
x_i^a : a^{th} symbol in column i

$$S(MSA) = \sum_{i=1}^{m} S(C_i)$$

$S(C_i)$: Score of column i
m: number of columns

f: Match $= 2$
Mismatch $= -1$
InDel $\quad = -2$

C	G	T	A	A
C	G	G	–	C
C	–	G	T	G
C	G	G	T	–

$S(C_1) = f(C,C) + f(C,C) + f(C,C) + f(C,C) + f(C,C) + f(C,C) = 12$

$S(C_2) = f(G,G) + f(G,-) + f(G,G) + f(G,-) + f(G,G) + f(-,G) = 0$

$S(C_3) = f(T,G) + f(T,G) + f(T,G) + f(G,G) + f(G,G) + f(G,G) = 3$

$S(C_4) = f(A,-) + f(A,T) + f(A,T) + f(-,T) + f(-,T) + f(T,T) \quad = -6$

$S(C_5) = f(A,C) + f(A,G) + f(A,-) + f(C,G) + f(C,-) + f(G,-) \quad = -9$

$S(MSA) = 12 + 0 + 3 - 6 - 9 = 0$

Fig. 5.6 A sample MSA and its score, S(MSA), using the sum-of-pairs approach. $f(-, -) = 0$

$$S(MSA) = \sum_{i=1}^{n-1} \sum_{j=i+1}^{n} S(Si, Sj)$$

n : number of sequences
$S(Si, Sj)$: pairwise alignment score between sequences Si and Sj.

Match   = 2
Mismatch = -1
InDel   = -2

S1	C	G	T	A	A
S2	C	G	G	–	C
S3	C	–	G	T	G
S4	C	G	G	T	–

$S(S1,S2) = 0$

S1	C	G	T	A	A
S2	C	G	G	–	C

$S(S2,S3) = -1$

S2	C	G	G	–	C
S3	C	–	G	T	G

$S(S1,S3) = -3$

S1	C	G	T	A	A
S3	C	–	G	T	G

$S(S2,S4) = 2$

S2	C	G	G	–	C
S4	C	G	G	T	–

$S(S1,S4) = 0$

S1	C	G	T	A	A
S4	C	G	G	T	–

$S(S1,S4) = 2$

S3	C	–	G	T	G
S4	C	G	G	T	–

$$S(MSA) = 0 - 3 + 0 - 1 + 2 + 2 = 0$$

Fig. 5.7 Score of the sample MSA shown in Fig. 5.6 calculated as the sum of induced pairwise alignments. S(MSA): Score of the MSA

this equivalency is demonstrated in Fig. 5.7. If induced pairwise alignments have two gap symbols aligned with each other, that column results in a score of zero.

5.3   Building

Given n sequences of length m, the number of possible MSAs is given by the formula [5]:

$$\sum_{i=m}^{mn} \sum_{j=0}^{i} (-1)^j \binom{i}{j} \binom{i-j}{i-m-j}^n$$

For $n = 5, m = 10$, this number is about 1.35×10^{38}. Therefore, finding the best MSA by scoring all possible MSAs is not feasible.

5.3.1   Exact Method

Defining the score of an MSA as the sum of its column scores lets us use the dynamic programming approach described for pairwise alignment. However, in this case, we do not have a matrix to fill but a hypercube.

To be more precise, if there are n sequences to be aligned, then every vertex of this hypercube in R^n represents the score of the alignment of the prefixes of the n sequences up until that point. Let's call the sequences to be aligned $S_1, \ldots, S_n$ and let $S(i, j)$ denote the subsequence of S between positions i and j, inclusive. Then, following our notation from Chap. 4, $W(i_1, \ldots, i_n)$ would denote the score of the best MSA for sequences $S_1(1, i_1), \ldots, S_n(1, i_n)$.

In the case of pairwise alignment $W(i, j)$ depended on three preceding, adjacent vertices (top, left, and diagonal). These vertices differed by the two coordinates of $W(i, j)$ by one, namely they represented $W(i - 1, j)$, $W(i, j - 1)$, and $W(i - 1, j - 1)$. In the case of MSA, since W has n coordinates, there are $2^n - 1$ vertices that "precede" it. In other words, in the recursive equation to calculate the value of $W(i_1, \ldots, i_n)$, we need to perform $2^n - 1$ operations, which implies a complexity of $O(2^n)$. For example, in the case of three sequences, we need to perform $2^3 - 1 = 7$ operations:

$$W[i, j, k] = max \begin{bmatrix} (W[i - 1, j - 1, k - 1] + f[S_1(i), S_2(j), S_3(k)] \\ W[i - 1, j - 1, k] + f[S_1(i), S_2(j), -] \\ W[i - 1, j, k - 1] + f[S_1(i), -, S_3(k)] \\ W[i - 1, j, k] + f[S_1(i), -, -] \\ W[i, j - 1, k - 1] + f[-, S_2(j), S_3(k)] \\ W[i, j - 1, k] + f[-, S_2(j), -] \\ W[i, j, k - 1] + f[-, -, S_3(k)] \end{bmatrix}$$

$f()$ can be calculated using the sum-of-pairs or the metric-based approaches. In case of the sum-of-pairs approach, we need Comb(n,2) operations which implies a complexity of $O(n^2)$.

Assuming the sequences are of equal length, m, in the pairwise alignment, we needed to fill a matrix of size m^2 to find the score of the best alignment. For dynamic programming approach to MSAs, number of cells or vertices that need to be calculated is m^n. Hence, the dynamic programming approach, i.e., the exact solution to the MSA problem for n sequences of equal length m has complexity $O(2^n n^2 m^n)$, which is not feasible. As the problem is NP-complete, we adopt heuristic solutions to the MSA problem.

5.3.2 Progressive

Before considering progressive methods for MSA, we need to define profile of an alignment and pairwise alignment of two profiles or a profile with another sequence. In Fig. 5.8, we show an example MSA and its corresponding profile. If an MSA has length m defined over an alphabet of size k, its profile is a $(k + 1) \times m$ matrix where each column corresponds to the ratio of the symbols that constitute the column.

Profile-profile alignments typically follow the pairwise alignment approach (dynamic programming with affine gap penalties) described in Chap. 4 ([6, 7]). When two columns of two MSAs are aligned, the score can be calculated in a variety of ways [8, 9], e.g., the dot-product or correlation between the frequency distribution of the symbols in the two columns. Nevertheless, often a profile sum-of-pairs approach is adopted where each pair of symbols that make-up the combined column is scored. For example, if we were to

MSA								
S1	C	–	T	A	A	T	A	–
S2	C	G	G	–	C	T	A	
S3	–	G	G	T	–	T	C	C
S4	C	G	G	T	T	T	C	–
S5	A	G	A	–	G	T	G	C
S6	A	A	–	C	–	T	G	C
S7	C	A	T	C	A	T	T	–
S8	C	–	G	A	C	T	T	C

Profile								
A	0.25	0.25	0.125	0.25	0.25	0	0.25	0
C	0.625	0	0	0.25	0.25	0	0.25	0.5
G	0	0.5	0.5	0	0.125	0	0.25	0
T	0	0	0.25	0.25	0.125	1.0	0.25	0
–	0.125	0.25	0.125	0.25	0.25	0	0	0.5

Fig. 5.8 A sample MSA of DNA sequences and its profile representation

align the second and eighth columns of the MSA in Fig. 5.8, we would calculate a score of $16xf(G, C) + 16xf(G, -) + 8xf(A, C) + 8xf(A, -) + 8xf(- - C)$ as $f(-, -)$ is zero. Often this paired score is scaled by the total number of possible pairs, n^2, which is 64 for this example.

The profile representation can directly be used to calculate the score of aligning two columns instead of finding each pair of aligned symbols. For example, continuing with the profile alignment of columns 2 and 8 in Fig. 5.8, the score would be $0.5x0.5xf(G, C) + 0.5x0.5xf(G, -) + 0.25x0.5xf(A, C) + 0.25x0.5xf(A, -) + 0.25x0.5xf(-, C)$. The coefficients represent the frequency of the symbols in their respective columns. This representation is equivalent to the scaled score calculated in the previous paragraph. Hence, the pairwise dynamic programming approach can easily be extended to aligning profiles.

It is also common to use sequence weights when scoring two columns resulting from the profiles [10, 11]. The weights are typically a function of the distance of the sequence in the profile to the average representation of the profile. The motivation behind using sequence weights is to minimize the bias that may be introduced by over-representation of a group of similar sequences in the MSA. Profile-profile alignment methods can readily be applied to profile-sequence alignments as a sequence can be represented by a profile.

Greedy

In the greedy MSA approach, first a distance between pairs of n sequences is calculated. Typically, the distance is based on the optimum pairwise global alignment between these sequences (e.g., its score or percent identity in the alignment) or k-mer frequencies. Then the two closest sequences are aligned into a profile. Now, the profile is treated as a new sequence and the distance is calculated for pairs of $(n - 1)$ sequences/profiles. Again, the pair with the minimum distance is combined into a profile.

Progressively, a sequence/profile pair is merged in exactly $(n - 1)$ steps until one final alignment is obtained. Like all progressive methods, the greedy approach is dependent on the order with which the sequence/profiles are merged. Merging the closest sequence/profile pair may not yield the best MSA as gaps that are formed in early steps are progressed to the final alignment.

Star

In star MSA, like the greedy approach, the distance between all pairs of sequences is calculated. The sequence that is on average most similar to all other sequences is chosen as the center sequence. Then each sequence is aligned with the center sequence progressively based on the order of their similarity to the center sequence. For n sequences, in steps 2 to $n - 1$, the sequences are progressively aligned to the profile MSA that has been formed thus far.

Clustal

The Clustal suite is the most popular progressive alignment method [10, 12–14]. The main flow of the Clustal approach is depicted in Fig. 5.9.

The Clustal method begins by calculating a distance between pairs of sequences. This was initially based on the pairwise alignment between the sequences. In later versions, a k-mer-based distance was used. In Fig. 5.9, a sample distance matrix among five sequences is shown. Then, a phylogenetic tree using the distance matrix is formed using the UPGMA algorithm. Earlier versions of the Clustal suite used the neighbour-joining algorithm to build

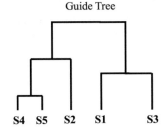

Distance Matrix

	S1	S2	S3	S4	S5
S1		.52	.21	.67	.72
S2			.56	.33	.28
S3				.48	.44
S4					.17
S5					

Guide Tree

S4   S5   S2   S1   S3

Order	Merged
1	S45=(S4,S5)
2	S13=(S1,S3)
3	S245=(S45,S2)
4	MSA=(S245,S13)

Fig. 5.9 The Clustal workflow. First, a distance matrix for the sequence pairs is obtained. Then, a phylogenetic tree based on the distance matrix is calculated. This tree is used as a guide to merge the sequences into profiles to form the final MSA

the tree. This tree is used as a guide to progressively merge sequences into profiles to form the final MSA.

In Fig. 5.9, the UPGMA tree based on the distance matrix suggests that the closest sequence pair is $S4$ and $S5$. Hence, first, these two sequences are merged into a profile. Next in the guide tree, following the linkage of the internal nodes, is the $S1 - S3$ sequence pair. After these two sequences are merged into a profile, $S2$ is merged with the $S4 - S5$ profile. The two remaining profiles of $S1 - S3$ and $S2 - S4 - S5$ are merged to obtain the final MSA. The order of the merge steps is depicted in Fig. 5.9.

During profile merges, earlier versions of the Clustal approach used sequence weighting based on the distance of the sequences to the root of the tree. The current version, Clustal Omega (Clustal O), uses hidden Markov model (HMM) based profile alignments during its progression. Also different from the earlier versions, Clustal O does not build a guide tree for all the sequences. First, the sequences are clustered using the k-means method. A guide tree is built for each cluster. These guide trees are merged using the UPGMA method and the final guide tree is used to progressively merge sequences into profiles and then to the final MSA using an HMM-based profile alignment method ([15].

GramAlign

GramAlign differs from most progressive MSA approaches both in the way it calculates distance between sequences and in the order the sequences are merged [16]. GramAlign uses a grammar-based distance measure for pairs of sequences [17]. This measure is based on Lempel-Ziv complexity [18], which is the number of steps used to produce a sequence using a "copy-paste + single letter addition" procedure. Each step uses the part of the sequence that has been produced thus far. For example, it takes 5 steps to produce the sequence "GGTACGTT" → "G - GT - A - C - GTT".

The distance used in GramAlign is based on the observation that if two sequences are similar, then it takes few numbers of steps to produce one from another. The ratio of the sum of the numbers of steps needed to produce one sequence from another to the complexity of the two sequences appended both ways is defined as the distance metric. This is represented as a network where the nodes are sequences, and the edges are the distance between sequences. A minimum spanning tree of this network is used as the order to merge the sequences into an MSA.

5.3.3 Iterative

One of the main problems with progressive approaches is the progression of errors. Incorrect alignments at early stages of the merge order are propagated to the final MSA. Iterative methods try to find a remedy for this problem by iteratively refining the progressive alignment phase. Popular iterative alignment methods include multiple alignment using fast Fourier transform (MAFFT) [19], multiple sequence comparison by log-expectation (MUSCLE) [7] profile alignment (PRALINE) [20], and MSAProbs [21]. Here, we will look in detail to two of these approaches.

MAFFT

MAFFT takes advantage of the fact that the FFT of the correlation function between two sequences when the second one is shifted by a lag of k is the product of the FFT of the sequences where the first one is complex conjugated. A high value of correlation with a lag of k can be interpreted as highly overlapping similarity when one of the sequences is shifted by k symbols and the two sequences are aligned without gaps. The ks for which this is true can quickly be identified and tested to find the actual overlapping regions of similarity.

MAFFT calculates correlation after each protein sequence is converted into a sequence of volume and polarity of its amino acids. For DNA/RNA sequences, the correlation is simply calculated for the symbol composition. These ideas are extended to correlation between groups of sequences where an average profile is obtained for the group. Once the highly similar regions between the sequences (or groups of sequences) are obtained, they are combined along the diagonal of the alignment matrix to define the final alignment.

The progressive alignment obtained this way is considered as a draft alignment, which is iteratively improved. The order (or the guide tree) with which the sequences are merged is divided into two subgroups that is different than the merge order [22]. New profiles and consequently a new MSA is built. This procedure is iteratively continued as long as the new MSAs show objective improvements.

MUSCLE

MUSCLE is divided into three stages. In the first stage a draft progressive MSA is built in a very similar way to the approach used in Clustal. The second stage is aimed to improve the guide tree and is the first of the two iterative approaches. The pairwise alignments that are deduced from the MSA obtained in the first stage are used to define the distance between pairs of sequences. This new distance matrix is used to build a new guide tree and consecutively a new progressive alignment. The second stage is terminated when there are no further significant changes to the guide tree structure.

The third stage iteratively improves the output of the second stage using a strategy similar to that of MAFFT. The guide tree of the second stage is converted into two disjoint sets by deleting an edge. The two resulting disjoint subsets of the guide tree represent two different profiles deduced from the MSA. These profiles are aligned to each other to obtain a new MSA. Edges are iteratively deleted and new MSAs are formed until a user-defined criteria (e.g., number of iterations, significant improvement) is achieved.

5.3.4 Constraint-based

Constraint-based approaches are further divided into two, based on the constraint being sequence or structure related. The first group is often referred to as consistency-based approaches and include popular methods such as tree-based consistency objective function for alignment evaluation (T_COFFEE) [23], ProbCons [24], and PSAlign [25]. The second group involves 2- or 3-D structure information of the sequences in building the MSA such Expresso [26] or PRALINE.

Consistency-based methods involve a library of alignment information that is used in building the MSA. Assume two regions of $S1$, Ra and Rb, both align well with a region of $S2$, Rc. Further assume that Rc aligns well with a region of $S3$, Rd, and this region of $S3$ (Rd) aligns well with Ra but not with Rb. Then, when determining whether to align Ra or Rb with Rc, the consistency-based algorithms choose to align Ra (not Rb) with Rc (since both Ra and Rc align well with Rd).

Structure-based approaches use known 2- or 3-D structures of parts of the sequences to supervise and refine the alignment process. Constraint-based approaches rely on external libraries, either in terms of global or local alignments or structure information, to guide the algorithm. This may induce additional space and time burdens but can result in improved accuracy.

5.4   Benchmarking

Collection of reference MSA databases such as BAliBASE (Benchmark Alignment dataBASE) [27] and HOMSTSRAD (HOMologous STRucture Alignment Database) [28] provide ground truth MSA data sets. These data sets are manually curated and refined to obtain final MSAs and represent cases that mimic real-life MSA problems. The data sets vary by size (number of sequences in the data set), length (average sequence length in the data set), similarity (the sequences in the data set may be highly-, somewhat-, or dis-similar), noise (fragmented or erroneous sequences may exist in the data set), and function (data sets may be intended to identify a domain, subfamily, motif, or evolutionary important signals).

MSA programs are tested by speed, space requirements, and their accuracy with respect to the reference MSAs. The accuracy is often assessed using two metrics. The first one is called the sum-of-pairs-score, which considers every pairwise alignment deduced from the MSA and counts the number of residues correctly aligned. The ratio of this sum for all pairs to the total number of pairwise residue alignments in the reference MSA constitutes the sum-of-pairs-score. The second metric is called the (total) column score, which is the ratio of the numbers of columns in the MSA that is identical to the columns in the reference MSA to the length of the reference MSA.

Other than manually curated reference MSAs, there is typically no single true MSA for a set of sequences. Often, the best practice is to use different MSA programs on augmented sequence sets (with sequences of known domains/motifs/functions) and manually inspect the final MSAs.

5.5   Exercises

1. Does a multiple sequence alignment (MSA) always induce a pairwise alignment of any two sequences in the MSA? Explain why or why not.

2. Calculate the "sum of pairs" score for the following multiple alignment. Use the following scoring scheme: Match = 2, Mismatch = -1, InDel = -2.

```
S1 G A T T G A − A −
S2 G A − T T A C A T
S3 T A C T T − C C T
S4 A A C − G A C A T
```

3. Describe the steps used in the Clustal algorithm to construct a multiple alignment of n sequences.

4. What is the main improvement iterative multiple sequence alignment approaches provide over the progressive approaches?

5. Assume we have pairwise sequence alignments of all pairs for n sequences, i.e., $n(n − 1)/2$ pairwise alignments in total. Would it always be possible to combine these pairwise alignments into a multiple sequence alignment such that the pairwise alignments are preserved. That is, for each pair of sequences, we want the induced pairwise alignment from the multiple sequence alignment be the same as the pairwise alignment in the starting collection. Explain why or why not.

6. NCBI's HomoloGene database catalogs groups of homologous genes. Obtain the protein sequences from group 87131, which represents the VEGFB gene, in FASTA format. You may open the FASTA file and shorten the sequence descriptions that are after the ">" symbol to just organism names for brevity.

A sister organization to NCBI is the European Molecular Biology Laboratory's (EMBL) European Bioinformatics Institute (EBI). Locate the multiple sequence alignment programs Clustal Omega, MAFFT, MAST, and T-Coffee on the EBI website. Perform multiple sequence alignment on the protein sequences obtained from HomoloGene using at least two of the four programs. Compare your results.

References

1. Terry Pratchett. *Thud!: A Discworld Novel*. Transworld Digital, 2007.
2. Pasi K Korhonen, Robin B Gasser, Guangxu Ma, Tao Wang, Andreas J Stroehlein, Neil D Young, Ching-Seng Ang, Deepani D Fernando, Hieng C Lu, Sara Taylor, et al. High-quality nuclear genome for sarcoptes scabiei-a critical resource for a neglected parasite. *PLoS Neglected Tropical Diseases*, 14(10):e0008720, 2020.

3. M. Vingron and P. R. Sibbald. Weighting in sequence space: a comparison of methods in terms of generalized sequences. *Proc Natl Acad Sci U S A*, 90(19):8777–81, 1993.

4. K. Sayood. *Introduction to Data Compression, Fifth Edition*. Morgan Kauffman-Elsevier, San Francisco, 2017.

5. J. B. Slowinski. The number of multiple alignments. *Mol Phylogenet Evol*, 10(2):264–6, 1998.

6. N. von Ohsen, I. Sommer, and R. Zimmer. Profile-profile alignment: a powerful tool for protein structure prediction. *Pac Symp Biocomput*, pages 252–63, 2003.

7. R. C. Edgar. Muscle: a multiple sequence alignment method with reduced time and space complexity. *BMC Bioinformatics*, 5:113, 2004.

8. G. Wang and Jr. Dunbrack, R. L. Scoring profile-to-profile sequence alignments. *Protein Sci*, 13(6):1612–26, 2004.

9. R. C. Edgar and K. Sjolander. A comparison of scoring functions for protein sequence profile alignment. *Bioinformatics*, 20(8):1301–8, 2004.

10. J. D. Thompson, D. G. Higgins, and T. J. Gibson. Clustal w: improving the sensitivity of progressive multiple sequence alignment through sequence weighting, position-specific gap penalties and weight matrix choice. *Nucleic Acids Res*, 22(22):4673–80, 1994.

11. S. Henikoff and J. G. Henikoff. Position-based sequence weights. *J Mol Biol*, 243(4):574–8, 1994.

12. D. G. Higgins and P. M. Sharp. Clustal: a package for performing multiple sequence alignment on a microcomputer. *Gene*, 73(1):237–44, 1988.

13. D. G. Higgins, A. J. Bleasby, and R. Fuchs. Clustal v: improved software for multiple sequence alignment. *Comput Appl Biosci*, 8(2):189–91, 1992.

14. F. Sievers and D. G. Higgins. Clustal omega. *Curr Protoc Bioinformatics*, 48:3 13 1–16, 2014.

15. M. Steinegger, M. Meier, M. Mirdita, H. Vohringer, S. J. Haunsberger, and J. Soding. Hh-suite3 for fast remote homology detection and deep protein annotation. *BMC Bioinformatics*, 20(1):473, 2019.

16. D. J. Russell, H. H. Otu, and K. Sayood. Grammar-based distance in progressive multiple sequence alignment. *BMC Bioinformatics*, 9:306, 2008.

17. H. H. Otu and K. Sayood. A new sequence distance measure for phylogenetic tree construction. *Bioinformatics*, 19(16):2122–30, 2003.

18. A. Lempel and J. Ziv. On the complexity of finite sequences. *IEEE Transactions on Information Theory*, IT-22:75 – 81, 1976.

19. K. Katoh, K. Misawa, K. Kuma, and T. Miyata. Mafft: a novel method for rapid multiple sequence alignment based on fast fourier transform. *Nucleic Acids Res*, 30(14):3059–66, 2002.

20. P. Bawono and J. Heringa. Praline: a versatile multiple sequence alignment toolkit. *Methods Mol Biol*, 1079:245–62, 2014.

21. J. Gonzalez-Dominguez. Fast and accurate multiple sequence alignment with msaprobs-mpi. *Methods Mol Biol*, 2231:39–47, 2021.

22. M. Hirosawa, Y. Totoki, M. Hoshida, and M. Ishikawa. Comprehensive study on iterative algorithms of multiple sequence alignment. *Comput Appl Biosci*, 11(1):13–8, 1995.

23. C. Notredame, D. G. Higgins, and J. Heringa. T-coffee: A novel method for fast and accurate multiple sequence alignment. *J Mol Biol*, 302(1):205–17, 2000.

24. C. B. Do, M. S. Mahabhashyam, M. Brudno, and S. Batzoglou. Probcons: Probabilistic consistency-based multiple sequence alignment. *Genome Res*, 15(2):330–40, 2005.

25. S. H. Sze, Y. Lu, and Q. Yang. A polynomial time solvable formulation of multiple sequence alignment. *J Comput Biol*, 13(2):309–19, 2006.

26. F. Armougom, S. Moretti, O. Poirot, S. Audic, P. Dumas, B. Schaeli, V. Keduas, and C. Notredame. Expresso: automatic incorporation of structural information in multiple sequence alignments using 3d-coffee. *Nucleic Acids Res*, 34(Web Server issue):W604–8, 2006.

27. A. Bahr, J. D. Thompson, J. C. Thierry, and O. Poch. Balibase (benchmark alignment database): enhancements for repeats, transmembrane sequences and circular permutations. *Nucleic Acids Res*, 29(1):323–6, 2001.

28. K. Mizuguchi, C. M. Deane, T. L. Blundell, and J. P. Overington. Homstrad: a database of protein structure alignments for homologous families. *Protein Sci*, 7(11):2469–71, 1998.

Molecular Phylogeny

6

In a nutshell ... Relationships are important, especially among organisms. We explore different ways of exploring relatedness based on the similarity of biological sequences.

6.1 Introduction

Phylogeny is the study of the evolutionary relatedness of organisms and groups of organisms. Traditionally evolutionary relationships were deduced from morphological and physiological similarity or dissimilarity. Thus mammals are evolutionarily related because they resemble each other physiologically in that they bear their young and provide nutrition through mammary glands. Beavers, hamsters, voles, capybaras, and rats are all rodents because they possess two pairs of rootless, continuously growing, incisors. The more similar the morphological features the closer the relationship is assumed to be.

While much can be inferred about evolutionary relationships by studying morphological and physiological similarities when dealing with mammals, sometimes these similarities can lead to confusing results. Consider the case of the giant panda. The panda looks like a bear but has a lot of features that are not particularly bearlike. While it is a carnivore, its diet consists mainly of bamboo shoots; the bear roars, the giant panda bleats; bears hibernate, giant pandas don't. The giant panda also shares similarities such as color patterning, skull structure, etc. with the red panda which for a long time was believed to belong to the raccoon (Procyonidae) family (it now constitutes a family of its own (Ailuridae)). So there was some question as to whether the giant panda really belonged to the bear family (Ursidae) or whether it should be classified as a member of the raccoon family. The papers published over a hundred-year period which examined this conundrum split almost evenly between bear and raccoon. What finally resolved the matter was the availability of a whole new set of features—molecular features—which could be used to answer the question of who was more closely related to whom.

© The Author(s), under exclusive license to Springer Nature Switzerland AG 2022
K. Sayood and H. H. Otu, *Bioinformatics*, Synthesis Lectures on Biomedical Engineering,
https://doi.org/10.1007/978-3-031-20017-5_6

In 1985 S.J. O'Brien and colleagues at the National Cancer Institute and the Smithsonian examined the similarity of DNA from the giant panda to DNA from the red panda, raccoon, American brown bear, spectacled bear, and the Malayan sun bear. They used several measures of similarity. They looked at the stability of the hybridization of single-stranded DNA from one population to single-stranded DNA from the other population. They looked at the relative distances traveled by similar proteins on an electrophoresis gel, and they looked at the strength of antigen-antibody reactions of homologous (more on this later) proteins. Each one of these techniques gave slightly different results for the relative closeness of the different species. However, on the central point, they all agreed. Turns out the giant panda is not closely related to the red panda. It is, however, very closely related to the bears. The red panda is not at all close to bears. Amongst these species, it is closest to the raccoon. The addition of molecular features finally resolved the hundred-year old disagreement.

When you are looking at mammals you can compare jawbones, or skulls, or other physical features. When comparing bacteria, the most abundant and diverse group of organisms on the planet, there are very few obvious characteristics we can use. We can tell if they look like rods or spheres. We can see how they react to staining—a process developed by Hans Christian Gram in 1884. We can see how different phages affect them. All this leads to a very crude classification. Clearly, we need additional features. Molecular biology provided the additional features we needed. With the availability of molecular information, evolutionary relationships can be inferred on a much finer scale by studying *homologous* genes in different organisms.

Homologous is a word we will see often so it is useful to define it. Two features are homologous if their similarity is due to shared ancestry. The wings of bats and the arms of humans are homologous because the common ancestor of bats and humans had structures that evolved into wings in bats and arms in humans. However, the wings of birds and the wings of bats may be considered to be homologous or not depending on the context. When considered as forearms they are homologous for the same reason that bat wings and human arms are homologous. Bats, birds, and humans all inherited their forearms from their last common ancestors. However, bat wings and bird wings evolved independently for flight so in the context of flight the two wings are not homologous. Instead, they are examples of homoplasy which is the convergent evolution of features responding to similar environmental selection. In molecular phylogeny, two genes in different organisms are said to be homologous if they are the direct descendants of a gene in the last common ancestor of the two organisms. There are two basic kinds of homology, *orthology* and *paralogy*. Two genes are said to be paralogous if they are descended from an ancestral gene that was copied in the genome, while in the case of orthologous genes the ancestral gene did not have copies. Thus, paralogous genes are homologous genes that reside in the same organism, while orthologous genes are homologous genes that reside in different organisms.

By counting the number of bases that differ between homologous genes, or the number of amino acids which differ between proteins we can get an idea of how much divergence there is between related species. Things get a bit trickier when we try to attach a timeline to the divergence. The nice thing about using jawbones and the like for establishing phylogeny is that often we also have ways of estimating the time at which species diverged through the fossil record. So how can we get an idea of how long ago two species diverged based on the differences between the nucleotide or amino acid sequences?

6.2 Molecular Clock

In 1962 Emile Zuckerkandl and Linus Pauling in an interesting paper [1] found a parallel between molecular disease and evolution; "... life is a molecular disease" is one of its observations. Of more interest to us is their examination of the hemoglobin molecule (specifically the α and β chains) obtained from different species. They found that the number of differences between evolutionarily close species, such as human and gorilla, are fewer than those between evolutionary distant groups. Even more interesting looking at the differences between the α chain between horse and human they assigned a time per each effective mutation—effective because we can only see the mutations that are present in the current-day descendants. They calculated a time of 11 to 18 million years per amino acid substitution. Given the very wide range of values for each mutation this is not a very precise timekeeping method. However, it provides some hope that we could use differences in commonly occurring proteins and the genes responsible for them to infer time since the divergence from the last common ancestor. Eric Margoliash studying the protein cytochrome C, while still providing all kinds of caveats, gave a more definitive time for each mutation in cytochrome C - 11 million years. As we can see from Table 6.1 this fits the data available to him at that time.

The idea that the number of differences between protein residues (and later nucleotides) could be used to assign a time in number of years since the last common ancestor, became known as the *molecular clock hypothesis* though neither Zuckerkandl and Pauling nor Margoliash had claimed this title. It turns out that this concept, while not strictly accurate, is useful as long as we know the limitations. These limitations were pointed out by Richard

Table 6.1 The number of different residues in Cytochrome C for various organisms and their accepted divergence (From E. Margoliash "Primary Structure and Evolution of Cytochrome C" Proceedings of the National Academy of Sciences, Vol. 50, 1963)

Species	Number of differences	Divergence in million years
Horse–Human	12	130
Horse–Pig	3	33
Pig–Chicken	10	108–150
Rabbit–Tuna	19	184–228
Rabbit–Yeast	45	465–520

Dickerson in a 1971 paper [2]. He noted that the relationship between the number of corrected differences between the same protein in different organisms and the elapsed time since their divergence could be different for different proteins, for different time periods, and for different taxa. The number of differences needs to be corrected because the only differences we see are those that are evident at this time. There might have been a number of changes at a particular location in the protein or nucleotide sequence but we can at most only see one change—the last one. Dickerson looked at three different proteins: cytochrome C, hemoglobin, and Fibrinopeptides. He found that the rate of corrected changes was almost constant for each of the three proteins, however, the rates for the three proteins were different.

Defining the *Unit Evolutionary Period* (UEP) as the time in millions of years (MY) for a one percent change between the same protein in two different organisms he calculated a UEP of 20 MY for cytochrome C, 5.8 MY for hemoglobin, and 1.1 MY for fibrinopeptides. The differences can be explained by looking at their functions. Proteins that have to interact with other proteins in order to fulfill their function will incur fewer changes in the regions where the interaction takes place. If the regions of interaction are large as in the case of cytochrome C the number of changes in the protein sequences will be fewer than if the regions of interaction are small as in the case of the fibrinopeptides. The change in rates in different taxa also makes sense. Organisms with shorter life spans such as mice will show more changes than organisms with longer life spans such as turtles. Mice will go through many more generations than turtles over the same period.

If we keep these limitations in mind the idea of a molecular clock can be very useful. For example, we could find a protein in Human and Chimpanzee from which we can determine the UEP. We can then use this UEP to determine the phylogenetic distance between Human and other primates using the same protein in the other primates.

6.3   Distance Measures

Given a set of orthologous genes, we can estimate their evolutionary relationship by comparing the aligned sequences. The simplest thing to do would be to compare the nucleotides in which they are different. However, this tends to undercount the number of differences that have occurred between these genes and their last common ancestor. In fact the plot of base pair differences versus time since divergence looks like a log plot. This is because as time goes by more and more of the mutations are likely to occur at positions where mutations have occurred previously. This will mean that the number of differences between orthologous genes will undercount the number of mutations that have occurred. Several different measures have been proposed that correct for this undercount. All of them rely on the process of point mutation to be a Poisson process. This means that the mutation process has to satisfy the following conditions. In the following we assume n_t is the number of mutations that have occurred in the gene by time t.

1. The number of mutations at time $t = 0$ is zero. This condition is trivially satisfied.
2. The number of mutations that occur in any interval of time is independent of the number of mutations that occur in any other disjoint interval.
3. The probability of k mutations occurring in any interval depends only on the size of the interval and not on where on the time axis the interval is located.
4. We can pick a unit of time such that the probability of two or more mutations occurring in this time interval is vanishingly small.

6.3.1 Jukes–Cantor

The Jukes–Cantor approach assumes that the probability of a mutation occurring at any nucleotide is the same regardless of the nucleotide and regardless of the mutation. Defining the transition probability matrix as

$$P_t = \begin{bmatrix} P_{AA} & P_{AC} & P_{AG} & P_{AT} \\ P_{CA} & P_{CC} & P_{CG} & P_{CT} \\ P_{GA} & P_{GC} & P_{GG} & P_{GT} \\ P_{TA} & P_{TC} & P_{TG} & P_{TT} \end{bmatrix}$$

where P_{XY} denotes the probability that a nucleotide at a particular location changes from X to Y in time t, in the Jukes–Cantor approach we pick

$$P_{XY} = \begin{cases} \alpha & \text{if } X \neq Y \\ 1 - 3\alpha & \text{if } X = Y \end{cases}$$

This selection results in the Jukes–Cantor distance

$$d = -\frac{3}{4} \ln \left(1 - \frac{4}{3} p \right)$$

where p is the proportion of nucleotides that are different in the two sequences.

Let's do a hand-waving derivation of this result. Let us define $q(t)$ to be the proportion of the nucleotides in the two sequences $\mathcal{A}$ and $\mathcal{B}$ that are the same. Then

$$q(t) = 1 - p(t)$$

At time $t + 1$ the proportion of nucleotides that remain the same belong to two groups: the nucleotides that were the same at time t and remain the same at time $t + 1$, and the nucleotides that were different at time t but are now the same. The proportion of nucleotides that were the same and stayed unchanged is given by the product of $q(t)$ with the probability of no change in those nucleotides in $\mathcal{A}$ and the probability of no change in those nucleotides in $\mathcal{B}$, or $q(t)(1 - 3\alpha)^2$. The nucleotides that were different and are now the same come about in one of three ways. The nucleotide in $\mathcal{A}$ stays the same and the nucleotide in $\mathcal{B}$ changes or the nucleotide in $\mathcal{B}$ stays the same, and the nucleotide in $\mathcal{A}$ changes, or they both change. Therefore,

$$q(t+1) = q(t)(1-3\alpha)^2 + (1-q(t))(\alpha(1-3\alpha) + \alpha(1-3\alpha) + \alpha^2)$$

Because α is small we can ignore terms involving powers of α and we get

$$q(t+1) = q(t)(1-6\alpha) + (1-q(t))2\alpha$$
$$= q(t) + 2\alpha - 8\alpha q(t)$$

or

$$q(t+1) - q(t) = 2\alpha - 8\alpha q(t)$$

As we make the interval smaller, in the limiting case we get

$$\frac{dq}{dt} = 2\alpha - 8\alpha q$$

Solving this differential equation we get the integrating factor

$$\mu = ke^{8\alpha t}$$

and the solution for $q(t)$ as

$$q(t) = \frac{1}{4} + ce^{-8\alpha t}$$

We can obtain the value of c by using the initial condition $q(0) = 1$. This follows from the assumption that we are looking at homologous genes which at the time of separation from the ancestral gene were identical. Thus

$$q(t) = \frac{1}{4} + \frac{3}{4}e^{-8\alpha t}$$

and the proportion of the nucleotides that are different is given by

$$p(t) = 1 - q(t)$$
$$= \frac{3}{4}\left(1 - e^{-8\alpha t}\right)$$

Given our initial probability assignments, the probability of change in a nucleotide per unit time is 3α. As this is true for both sequences the distance d for a period t will be $6\alpha t$. From above we have

$$e^{-8\alpha t} = 1 - \frac{4}{3}p$$

or

$$-8\alpha t = \ln(1 - \frac{4}{3}p)$$

Substituting $d = 6\alpha t$ we get

$$d = -\frac{3}{4}\ln(1 - \frac{4}{3}p)$$

6.3.2 Kimura 2 Parameter

In order to obtain the Jukes–Cantor distance we assumed that the probability of changing from a base to any other base was the same. However, we know that this is not true in reality; the probability of changing from a purine to a purine or pyrimidine to a pyrimidine (transition) is higher than the probability of going from a purine to a pyrimidine and vice versa (transversion). The Kimura 2 parameter distance takes this into account by assuming different probabilities for transitions and transversions. In this model the transition probability matrix becomes

$$P_t = \begin{bmatrix} 1 - 2\beta - \alpha & \beta & \alpha & \beta \\ \beta & 1 - 2\beta - \alpha & \beta & \alpha \\ \alpha & \beta & 1 - 2\beta - \alpha & \beta \\ \beta & \alpha & \beta & 1 - 2\beta - \alpha \end{bmatrix}$$

The total frequency of occurrence of transition pairs P, can be obtained as

$$P = \frac{1}{4}\left(1 - 2e^{-4(\alpha+\beta)t} + e^{-8\beta t}\right)$$

and the total frequency of transversion pairs Q can be obtained as

$$Q = \frac{1}{2}\left(1 - e^{-8\beta t}\right)$$

The evolutionary distance d can be expressed in terms of observed P and Q as

$$d = \frac{1}{2}\ln\left[\frac{1}{1 - 2P - Q}\right] + \frac{1}{4}\ln\left[\frac{1}{1 - 2Q}\right]$$

6.4 Alignment Free Distances

Calculating the Jukes–Cantor and Kimura 2 parameter distances requires aligning the sequences. In the case of multiple sequences, this would mean a multiple alignment. Multiple alignments can be tricky and the results can depend on the selections made for the various penalties. Therefore, there is interest in developing distance measures that do not require using alignment. The problem with these measures is that there is no biological justification for them. However, they seem to provide valid phylogenies and they are much faster than alignment based methods.

Several alignment-free methods use the concept of Kolmogorov complexity. The development of the idea of Kolmogorov complexity can be attributed to three individuals, Ray Solomonoff, Andrey Kolmogorov, and Gregory Chaitin (the latter when he was a high school student). Ray Solomonoff has the priority of discovery but his work was not widely known and the same concepts were independently discovered by Kolmogorov and Chaitin. As Kolmogorov was the better known of the two, by the Matthew effect, the idea became known as Kolmogorov complexity; Chaitin always described it as Algorithmic complexity—a more descriptive name that leaves the issue of discovery alone. The Kolmogorov complexity of a sequence S, denoted by $K(S)$ is defined as the length of the shortest program that can generate the sequence. We can also define the conditional Kolmogorov complexity of a sequence Q given that we know S, denoted by $K(Q|S)$ as the length of the shortest program which could generate Q using the knowledge of S. We can define a distance between sequences Q and S as some function of the conditional Kolmogorov complexity. Consider the two extreme cases where Q is unrelated to S and where Q is the same as S. In the first case, knowledge of S provides no advantage to the program generating Q therefore, the sizes of the shortest programs with and without the knowledge of S will be the same. Therefore, the conditional Kolmogorov complexity $K(Q|S)$ will be the same as the Kolmogorov complexity $K(Q)$. In the second case, the program generating Q will simply write out S so the size of the program, or $K(Q|S)$ is effectively zero. Normalizing this conditional complexity to take out the effect of the length of Q we can obtain a useful distance measure. While the idea of using the Kolmogorov complexity in this fashion is a good one there is one slight problem. Nobody knows how to compute the Kolmogorov complexity of an arbitrary sequence. However, we can try and approximate the Kolmogorov complexity and there are several suggestions on how to do this. We will look at one of them.

In a paper in 1976 Abraham Lempel and Jacob Ziv proposed a method for computing the complexity of a sequence which became the basis for several popular lossless compression schemes. We will refer to the complexity measure as LZ (for Lempel-Ziv) complexity and use it to approximate Kolmogorov complexity. First, some basic nomenclature: Suppose we have sequences S, Q, and R. Define l_S to be the length of S, S_i to be the ith element in the sequence S, and $S_{i,j}$ to be the subsequence of S $\{S_i, S_{i+1}, \ldots, S_j\}$. Suppose R is a concatenation of S with Q which we denote by $R = S \cdot Q$. R is said to be *reproducible* from S (denoted by $S \rightarrow R$) if Q can be generated by copying starting from some R_i where $i < l_S$. That is a somewhat weird way of putting it. Why not just say that $S \rightarrow R$ if Q is a subsequence of S? We can see why by looking at a couple of examples. First the obvious case: suppose $S = ACACG$ and $Q = CACG$ then we can generate $R = ACACGCACG$ by beginning with S and copying the second through the last element of S. Clearly $S \rightarrow R$. Now consider the case where we have the same S but $Q = CGCG$. At first glance, it seems we cannot say that $S \rightarrow R$ because $CGCG$ is not a subsequence of S. But look again. Suppose we start copying from the fourth element of S and continue copying. After the first copy we get $ACA\underline{C}GC\ldots$ where we have underlined the letter we have copied. We then copy the next letter G to get $ACAC\underline{G}CG \ldots$. In this new sequence we still have more letters to copy so we copy the next letter to get $ACACGC\underline{G}C\ldots$. We can now copy the next letter $ACACGC\underline{G}CG$ to obtain R. Even though Q was not a subsequence of S, $R = S \cdot Q$ is still reproducible from S. In other words we still have $S \to S \cdot Q$. If we add one more degree of freedom to this process we get the operation of *producibility*. We say one sequence is producible from another if all but the last element of the latter sequence is reproducible from the former sequence. Consider S as defined earlier and $Q = CACT$. If $R = S \cdot Q = ACACGCACT$, then we can see that all but the last letter of the sequence R is reproducible from the sequence S. If a sequence R is producible from S we denote this by $S \Rightarrow R$.

We can use the concept of producibility to produce a parsing of a sequence. We start from an empty sequence and start trying to generate the sequence in question. At each step the portion we extend the sequence being constructed becomes an element in the parsing. The number of elements in the parsing gives us an estimate of the Kolmogorov complexity. For example, consider the sequence $S = ACACGCGACGCT$. Starting from the empty sequence ϕ we can generate S in the following steps.

$$\phi \Rightarrow A$$

$$A \Rightarrow AC$$

$$AC \Rightarrow ACACG$$

$$ACACG \Rightarrow ACACGCGA$$

$$ACACGCGA \Rightarrow ACACGCGACGCT$$

We can obtain a parsing of the sequence S as $A \cdot C \cdot ACG \cdot CGA \cdot CGCT$ where each new element is what was added on in a producing step. Because the number of elements in the parsing is five we say that the complexity of the sequences is five. To differentiate this definition from Kolmogorov complexity we will refer to this measure of complexity as Lempel-Ziv complexity or LZ complexity and denote it by $c(\cdot)$.

Let's use this measure of complexity to obtain a measure of distance. We can approximate $K(\cdot)$ with $c(\cdot)$, but what about the conditional Kolmogorov complexity $K(Q|S)$? The conditional complexity is the complexity of a sequence Q when another sequence S is already available. In terms of LZ complexity, this would be like having the sequence S available for copying when parsing sequence Q. The number of elements in the parsing of Q given S was available would be given by $c(S \cdot Q) - c(S)$.

We will see how we can use this measure of complexity to obtain a measure of distance between two sequences through the use of an example. Let's suppose S is as has been defined previously and we have two other sequences $Q = GAGACGCTGAGT$ and $R = CTCTGCTCTCACA$. We will compute a measure of similarity or distance between S and Q and between S and R. We will do so using LZ complexity in place of Kolmogorov complexity. In order to compute the distance we need $c(Q)$, $c(R)$, $c(SQ)$, and $c(SR)$. We compute these as follows:

$$Q = G \cdot A \cdot GAC \cdot GC \cdot T \cdot GAGT \ \text{ therefore, } c(Q) = 6$$

$$R = C \cdot T \cdot CTG \cdot CTCTA \cdot CA \ \text{ therefore } c(R) = 5$$

$$S \cdot Q = A \cdot C \cdot ACG \cdot CGA \cdot CGCT \cdot GAG \cdot ACGCTGAGT \ \text{ therefore } c(S \cdot Q) = 7$$

$$S \cdot R = A \cdot C \cdot ACG \cdot CGA \cdot CGCT \cdot CTCTG \cdot CTCTCA \cdot CA \ \text{ therefore, } c(S \cdot R) = 8$$

and

$$c(S \cdot Q) - c(Q) = 7 - 6 = 1; \quad c(S \cdot R) - c(R) = 8 - 5 = 3$$

Thus, though the complexity of Q is greater than the complexity of R, the conditional complexity of Q given S is less than the conditional complexity of R given S. In order to convert this measure of similarity into a valid distance measure we need to make sure the measure follows the rules for distance metrics. The following function satisfies these requirements:

$$d(S, Q) = \frac{max\{c(SQ) - c(S), c(QS) - c(Q)\}}{max\{c(S), c(Q)\}}$$

and we can use it to obtain the distance between sequences. The advantage of this approach to finding distances is that it does not require the computationally expensive process of alignment, it does not require the estimation of parameters such as the gap opening and extension penalties, and it is not based on any particular mathematical model of evolution. The disadvantage is that it is not based on any particular mathematical model of evolution. This lack of an underlying model of evolution means that interpretations of distances computed in this manner rely on correlating the distances generated with existing knowledge about evolutionary distances between species.

6.5   Tree Building

Once we have obtained the distance between various sequences we need to represent their evolutionary relationships. We do so in the form of a tree. Before we go into the details of how we go about building trees, let us briefly look at some of the nomenclatures. In order to do this let's use the made-up example shown in Fig. 6.1. If there is a node to which all branches converge this is called a root node. The tree on the left in Fig. 6.1 is a rooted tree while the tree on the right is not. A rooted tree is obtained if we postulate a common ancestor for all the organisms under consideration. This seems simpler than it is. What we have available to us is a set of pairwise differences. This allows us to build relationships between organisms but it does not necessarily specify the root.

The terminating point on the tree is usually called a leaf. The branches connect the nodes (how is that for a circular definition) or connect a node to the terminating point. As we progress from the root to the leaves in a rooted tree we are tracing evolutionary development. When we look at a pair of connected nodes, the organism or group of organisms on the root

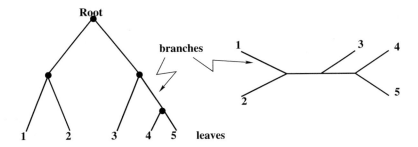

Fig. 6.1 Trees

side of the tree are the proposed or inferred most recent common ancestor of the organism or organisms on the leaf side of the tree. Usually the tree is a binary tree. This is not much of a restriction as most trees can be represented within some degree of fidelity as a binary tree.

Let's see how we can obtain a tree using an example. Suppose we have five species, A, B, C, D, and E with distance matrix:

	A	B	C	D	E
A		4	8	8	6
B			8	8	6
C				4	6
D					6
E					

A possible tree that agrees with these distances is shown in Fig. 6.2. However, it is not clear where the root of the tree should be. To obtain a rooted tree the usual approach is to specify an *outgroup* species whose distance from all other species under consideration is greater than the distance between individual species. Thus the common ancestor of the outgroup species and the rest of the species forms the root of the tree as shown in Fig. 6.3.

There are several different ways of classifying tree-building approaches. We can classify them as clustering approaches or search approaches, where the search is for a tree that

Fig. 6.2 A possible tree

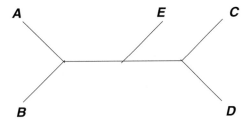

Fig. 6.3 Rooting a tree

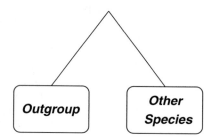

satisfies an optimality criterion. We can also classify them as distance-based approaches where the sequence relationships are abstracted in terms of distances between them, and character-based approaches in which we deal with the actual sequences themselves. There are also sub-tree methods in which smaller trees are combined to form a larger tree. We look at these approaches in turn.

6.5.1   Distance-Based Methods

As the name implies distance-based methods use the distances calculated between species to form the tree.

The two most commonly used distance-based algorithms are the Unweighted Pair Group Method Using Arithmetic Averages (UPGMA) and Neighbor Joining.

UPGMA

UPGMA is essentially an unsupervised hierarchical clustering algorithm that is easy to use but may not be entirely suited for phylogenetic applications. However, its ease of use has made it a popular method at least for getting a first approximation to the final tree. The basic approach is as follows: Given a set of pairwise distances find the closest pair (if there are more than one closest pair then pick one of them at random) and treat them as the leaves of a common node. The common node is now treated as a leaf and the distances between this new leaf and the rest of the leaves are now computed. There are several different ways of computing these distances leading to different variants of the UPGMA algorithm - more on this later. Again find the smallest pairwise distance and continue. Repeat until the tree is complete.

The most common way the distances between the new leaf and the other leaves in the tree are computed is called average linkage clustering. As the name implies the new distances between the common node and the other leaves in the tree are computed as the average of the distance between the offsprings of the common node and the other leaves. Let's suppose the closest pair of leaves are leaves labeled i and j and they are now the offsprings of the common node v. Let's suppose i has n_i offsprings and j has n_j offsprings. Then the distance between v and another leaf k are computed as

$$d(v, k) = \frac{n_i}{n_i + n_j} d(i, k) + \frac{n_j}{n_i + n_j} d(j, k)$$

which is known as average linkage clustering. If instead of using the average distance we use the minimum distance as in

$$d(v, k) = \min_{i,j} \{d(i, k), d(j, k)\}$$

this is called single linkage clustering. If we use the maximum distance of the offsprings

$$d(v, k) = \max_{i,j} \{d(i, k), d(j, k)\}$$

this is called complete linkage clustering.

To see how the average linkage UPGMA algorithm works let's use the distances between the 12S rRNA genes from Human, Gibbon, Gorilla, Orangutan, and Chimpanzee. The following distances were calculated using the complexity method.

Human	0.000	0.584	0.349	0.491	0.344
Gibbon	0.584	0.000	0.592	0.632	0.577
Gorilla	0.349	0.592	0.000	0.502	0.357
Orangutan	0.491	0.632	0.502	0.000	0.548
Chimpanzee	0.344	0.577	0.357	0.548	0.000

Looking through the table we find that the smallest distance is between the Human and Chimpanzee genes so we assume these are the leaves of a common node. Let's call the common node HuCh and calculate the distances of all other species to this common node. For example, the distance between Gibbon and HuCh would be the average of the distance between Human and Gibbon and Chimpanzee and Gibbon or $\frac{0.584+0.577}{2} = 0.581$. Similarly, computing the other distances we get the following distance table:

HuCh	0.000	0.581	0.353	0.520
Gibbon	0.581	0.000	0.592	0.632
Gorilla	0.353	0.592	0.000	0.502
Orangutan	0.520	0.632	0.502	0.000

The smallest distance in the table is now between HuCh and Gorilla, so we assign HuCh and Gorilla as offsprings of a common node which we call HuChGo. The distance between HuChGo and Gibbon is given by

$$d(HuChGo, Gi) = \frac{2}{3}d(HuCh, Gi) + \frac{1}{3}d(Go, Gi)$$
$$= \frac{2}{3} \times 0.581 + \frac{1}{3} \times 0.592$$
$$= 0.585$$

Similarly, the distance between HuChGo and Orangutan will be

$$d(HuChGo, Or) = \frac{2}{3} \times 0.520 + \frac{1}{3} \times 0.502 = 0.514$$

The new distance table is:

HuChGo	0.000	0.585	0.514
Gibbon	0.585	0.000	0.632
Orangutan	0.514	0.632	0.000

Now the smallest distance is between HuChGo and Orangutan which dictates the next joining of the tree to generate the node HuChGoOr. The node HuChGo has three offsprings so the distance between HuChGoOr and Gibbon is given by

$$d(HuChGoOr, Gi) = \frac{3}{4} \times 0.585 + \frac{1}{4} \times 0.632 = 0.597$$

This node is then combined with Gibbon resulting tree is shown in Fig. 6.4.

Notice that regardless of the way we measure distance, the distance of the common node to each of its leaves is the same. This means that the distance from the root node to each of the leaves is the same. This kind of tree is called an *ultrametric* tree. If we treat the distance of a leaf from a node as a measure of time we can infer a molecular clock that tells us how long it has been since the arising of species from a common ancestor. However, this implies that the rate of evolution for each species was the same which is not necessarily true.

Fig. 6.4 A tree full of primates

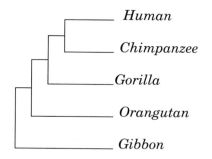

Neighbor-Joining

The Neighbor-Joining algorithm proposed by Saitou and Nei in 1987 is one of the most popular methods of tree building. The approach begins with a star topology with all leaf nodes coming out of the same node. The star topology is used to construct a binary tree by successively joining nodes into pairs of nodes arising from a common parent. Such pairs of nodes are called neighbors. The basic idea is to pick neighbors that will result in the smallest sum of branch lengths. The reason for minimizing the sum of branch lengths is that it can be shown that for distance metrics with certain desirable properties the tree with the smallest sum of branch lengths will correspond to the minimum evolution tree. We explain the method using an example from Saitou and Nei's paper.

We begin with all nodes arising out of a common node X as shown in Fig. 6.5. We denote the distance between leaf nodes i and j by D_{ij} and the distance between two nodes a and b where one or both of the nodes are not leaf nodes by L_{ab}. Given the tree shown in Fig. 6.5 the sum of the branch lengths is given by

$$S_0 = \sum_{i=1}^{N} L_{iX}$$

We can also write the sum of the branch lengths in terms of the distances between the leaf nodes as

$$S_0 = \frac{D_{12} + D_{13} + \cdots + D_{1N} + D_{23} + \ldots D_{2N} + \ldots D_{(N-1)N}}{N-1}$$

or

$$S_0 = \frac{1}{N-1} \sum_{i=1}^{N-1} \sum_{j=i+1}^{N} D_{ij} = \frac{1}{N-1} \sum_{i<j} D_{ij}$$

To see why we need to normalize by $N-1$ consider the case where $N = 4$. If we write the un-normalized sum of the distances D_{ij} in terms of L_{iX} we get

$$D_{12} + D_{13} + D_{14} + D_{23} + D_{24} + D_{34}$$
$$= L_{1X} + L_{2X} + L_{1X} + L_{3X} + L_{1X} + L_{4X} + L_{2X} + L_{3X} + L_{2X} + L_{4X} + L_{3X} + L_{4X}$$
$$= 3L_{1X} + 3L_{2X} + 3L_{4X}$$

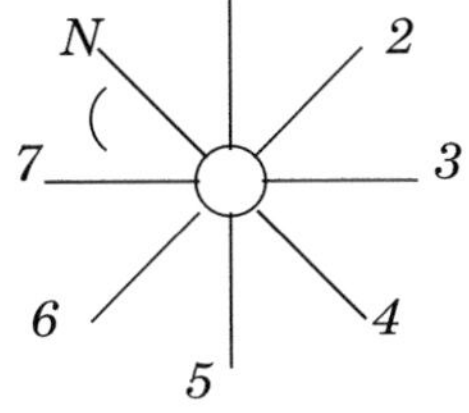

Fig. 6.5 Star topology

Fig. 6.6 Stage 1 of the
neighbor joining algorithm

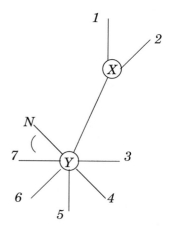

where $D_{ij} = L_{iX} + L_{jX}$. Because L_{iX} has been repeated three times for each i we need to divide by 3 (or $N - 1$) in order to obtain the sum of the branches.

Suppose we select nodes 1 and 2 to be the neighbors and assign them as arising from a single node which is then connected to the other nodes which are in a star topology as shown in Fig. 6.6.

We would like to find the sum of the branch lengths in this tree, denoted by $S_{[12]}$.

$$S_{[12]} = L_{XY} + (L_{1X} + L_{2X}) + \sum_{k=3}^{N} L_{kY}$$

While the sum is written in terms of L_{ab} what we have available is the set of distances $\{D_{ij}\}$ so we need to write L_{XY}, L_{iX}, and L_{iY} in terms of D_{ij}. In this tree L_{XY}, the distance between the nodes X and Y, can be computed by first computing the sum of the distances of nodes 1 and 2 to all other nodes in the tree and then subtracting from it the distances of the leaf nodes to the nodes X and Y.

$$L_{XY} = \frac{1}{2(N-2)} \sum_{k=3}^{N} (D_{1k} + D_{2k}) - \frac{1}{2}(L_{1X} + L_{2X}) - \frac{1}{N-2} \sum_{k=3}^{N} L_{kY}$$

Noting that

$$L_{1X} + L_{2X} = D_{12}$$

and

$$\sum_{k=3}^{N} L_{kY} = \frac{1}{N-3} \sum_{3 \le i < j} D_{ij}$$

we can write L_{XY} in terms of the distances D_{ij}.

Substituting the expression for L_{XY} into the expression for $S_{[12]}$ we get

$$S_{[12]} = \frac{1}{2(N-2)} \sum_{k=3}^{N} (D_{1k} + D_{2k}) - \frac{1}{2}(L_{1X} + L_{2X})$$

$$-\frac{1}{N-2} \sum_{k=3}^{N} L_{kY} + (L_{1X} + L_{2X}) + \frac{1}{N-3} \sum_{3 \le i < j} D_{ij}$$

$$= \frac{1}{2(N-2)} \sum_{k=3}^{N} (D_{1k} + D_{2k}) + \frac{1}{2}(L_{1X} + L_{2X}) + \frac{N-3}{N-2} \sum_{k=3}^{N} L_{kY}$$

$$= \frac{1}{2(N-2)} \sum_{k=3}^{N} (D_{1k} + D_{2k}) + \frac{1}{2}D_{12} + \frac{1}{N-2} \sum_{3 \le i < j} D_{ij}$$

Minimizing $S_{[12]}$ is equivalent to minimizing $a(S_{[12]} - b)$ as long as $a > 0$, and a and b are constants, or do not depend on the pair of nodes, 1 and 2 in this case. Let's take $a = 2$ and

$$b = \frac{1}{N-2} \sum_{i=1}^{N-1} \sum_{j=i+1}^{N} D_{ij} = \frac{1}{N-2} \sum_{1 \le i < j} D_{ij}$$

the sum of pairwise distances between the nodes scaled by $N - 2$. Let's define $S_{[12]}^* = 2(S_{[12]} - b)$. Then,

$$S_{[12]}^* = D_{12} + \frac{1}{N-2} \left(\sum_{k=3}^{N} D_{2k} + \sum_{3 \le i < j} D_{ij} - \sum_{1 \le i < j} D_{ij} \right)$$

$$+ \frac{1}{N-2} \left(\sum_{k=3}^{N} D_{1k} + \sum_{3 \le i < j} D_{ij} - \sum_{1 \le i < j} D_{ij} \right)$$

$$= D_{12} - \frac{1}{N-2} \sum_{k \ne 1} D_{1k} - \frac{1}{N-2} \sum_{k \ne 1} D_{2k}$$

Let's define

$$u_i = \frac{1}{N-2} \sum_{i \ne k} D_{ik}$$

as the "almost" average distance of node i to all other nodes. Then,

$$S_{[12]}^* = D_{12} - u_1 - u_2$$

or, in general

$$S_{[ij]}^* = D_{ij} - u_i - u_j$$

Computing $S_{[ij]}^*$ for all pairs of leaf nodes we find the pair (k, l) for which this quantity is the smallest. Note that to minimize $S_{[ij]}^*$ we need to keep D_{ij} small, that is we need a pair of leaves close to each other, and we need u_i and u_j large, which will be the case when this pair is far from everything else. This is a better reflection of reality than the case of UPGMA where we combine the closest pair at each iteration. Also note that finding the pair (k, l) that minimizes $S^*_{[ij]}$ also minimizes $S_{[ij]}$ which results in the smallest sum of branch lengths.

The distances of the leaf nodes k and l to the merged node $[kl]$ are given by

$$D_{k[kl]} = \frac{1}{2}(D_{kl} + u_k - u_l)$$

$$D_{l[kl]} = \frac{1}{2}(D_{kl} + u_l - u_k)$$

That is, we split the distance between k and l such that the leaf node that is on average further away from all other nodes is placed further away from the merged node. This can also be seen as an improvement over UPGMA as such a split does not result in an ultrametric tree and hence does not require the assumption of identical rates of evolution.

We can now update our distance matrix using

$$D_{[kl]j} = \frac{1}{2}(D_{kj} + D_{lj} - D_{kl}), \qquad j \neq k, l$$

The process is repeated until we are left with three nodes. The tree is iteratively drawn at each step as L_{XY} can be calculated using the elements of the distance matrix at each iteration.

The UPGMA and Neighbor-Joining methods are based on a measure of distance between sequences. However, methods of constructing phylogeny long precede the sequencing of genomes. There are several trait based techniques that were widely used to obtain phylogenetic relationships before sequencing of genomes became possible. These have now been updated to work with sequence-based "traits".

6.5.2   Method of Parsimony

In the method of parsimony trees are evaluated based on the idea that the best tree is the one that requires the least amount of evolution. Maximum parsimony methods existed before the advent of molecular phylogeny and have been adapted to use gene sequences rather than physical or morphological traits. Where the characters used to determine the tree were physical characteristics, such as color of eyes, number of teeth, etc., we now use the bases in a gene or amino acids in a protein as characters.

Given a set of aligned sequences we usually ignore all positions in the alignment that are in complete agreement—that is, each base in that location is the same. We usually also ignore all positions where there is complete disagreement—that is each base in that position is unique. Unlike the distance methods, a tree is evaluated on a single nucleotide from each of the sequences being examined. A candidate tree is evaluated by generating the ancestor sequence. The number of mutations required to generate the sequences being considered as directed by the tree is obtained. The tree from which the sequences being evaluated can be obtained with the smallest number of mutations is selected.

The procedure begins with a multiple sequence alignment. The locations where all the bases are the same are ignored and we focus on those locations where there is a difference in the bases.

Let's use an example to work through the process. Let's suppose we have four organisms for which we are trying to find a tree and we are given the following aligned set of "genes" to find the tree:

Organism	Sequence
a	G A G A C A T
b	G G G A A A T
c	G A G C T A T
d	G G G G G G T

In this alignment columns 1, 3, and 7 are uninformative as all the bases in this position are the same. Column 5 also does not contain any useful information as the bases are all different. Therefore, we will select our tree based on columns 2, 4, and 6. With four taxa we have three possible topologies as shown in Fig. 6.7.

We try out all three topologies for the three locations. In order to determine the number of mutations required we need to specify the ancestor base at two locations. The various possibilities are shown in Fig. 6.8 with the required mutations marked on the corresponding branches. There are two possibilities using the first topology which will give us a total number of mutations, or *tree length* of 2. We can now continue with this same topology and the bases in columns 4 and 6. Once we are done evaluating these trees we then evaluate the trees corresponding to the second topology and then the trees corresponding to the third topology. Even for this small toy example, the number of possible trees that need

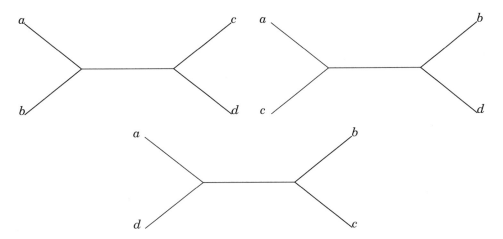

Fig. 6.7 Three possible topologies for four taxon

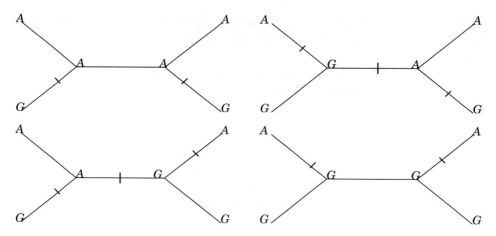

Fig. 6.8 Various possibilities for the first topology and the bases in the second location

to be evaluated is quite large. For any reasonable number of taxon and sequence size, the computational load for generating all possible trees soon becomes prohibitive.

The number of binary trees with k nodes is given by the Catalan number C_k which is given by

$$C_k = \frac{1}{k+1} C_k^{2k} = \frac{2k!}{(k+1)!k!}$$

The number of nodes k in a binary tree with n leaves is $2n - 1$. Therefore the number of binary trees for n species increases rapidly with n. If the number of binary trees is not unreasonably large we can evaluate each of the trees using the Fitch algorithm [3] which can find the length of the tree in linear time. If the number of trees is going to become unreasonably large we can use a branch-and-bound algorithm that restricts the number of trees that are generated. Let's look at each method in turn.

The Fitch Algorithm

The Fitch algorithm first proceeds from the leaves to the root of a rooted tree generating possible labels at each node as well as the tree length. At each node, we examine the traits—the sequences in this case—associated with the offsprings of the node. If we compare the set of labels at each of the offsprings, there are two possibilities; the intersection of the two sets is empty, or it is not. If the intersection is empty we label the node with the union of the sets of sequences of the offsprings and increment the tree length by one. If the intersection is not empty we label the node with the intersection. Once we reach the root we will have the tree length of this particular topology. We can also generate the actual tree by moving from the root down to the roots and in the process select the letter at each node from the sequences generated in the leaves-to-root progression. Depending on how we make the selections we

Fig. 6.9 Candidate tree

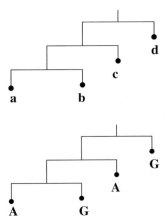

Fig. 6.10 The tree to be evaluated with the leaves labeled with the bases in column two

can end up with different assignments for the consensus sequence, however, we always get the same tree length.

It is a lot easier to see how this work using an example. We will use the example in the previous section. Let's pick the tree shown in Fig. 6.9 as the candidate tree.

We label the leaves of the tree with the letters in the second column of the alignment (See Fig. 6.10).

Looking at the lowest leaves we can see that one is marked with an *A* and the other with a *G*. There is no overlap between the two sets so we label the parent node with $\{A, G\}$ and increment the tree length by one (See Fig. 6.11).

The intersection between the sets $\{A, G\}$ and $\{A\}$ is $\{A\}$ so we label the next node up with *A* and do not increment the tree length (See Fig. 6.12).

Fig. 6.11 Assign union of $\{A\}$ and $\{G\}$ to the parent node and increment tree length by one

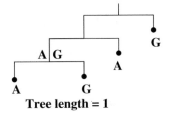

Fig. 6.12 Label node with $\{A, G\} \bigcap \{A\} = A$

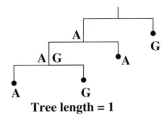

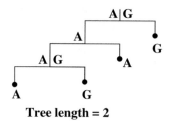

Tree length = 2

Fig. 6.13 Label root of the tree with the $\{A\} \bigcup \{G\} = \{A, G\}$ and increment tree length by one

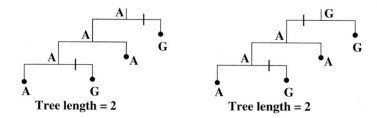

Tree length = 2   **Tree length = 2**

Fig. 6.14 The two trees obtained when traversing from the root to the leaves. The mutations are shown by vertical bars

At the root the descendants have labels $\{A\}$ and $\{G\}$ so the we label the root with the union $\{A, G\}$ and increment the tree length by one (See Fig. 6.13).

Once we have reached the root of the tree we can reverse direction and go from the root to the leaves. At each step, if a node is labeled with more than one letter we pick the letter which results in the least number of mutations. In this particular example, at the root we could pick either A or G. Let's pick A at the root node. In this case to get the right offspring will require a mutation as it is a G. The left offspring is an A so no mutation is required. From here the right offspring is an A so no mutation is required. The left offspring is the set $\{A, G\}$. In order to keep the number of mutations low we pick A. Instead of picking an A at the root node we could have picked a G. We would again have ended up with a tree requiring two mutations. The two trees are shown in Fig. 6.14.

Branch and Bound

While the Fitch algorithm gives us an efficient way of evaluating a candidate tree the exponential growth in the number of possible trees is still a problem. This makes it very difficult to evaluate trees exhaustively. One approach to reducing the number of trees to be explored is the branch and bound approach. In this approach, we determine a bound which is the largest a tree length can be. One way to do this is to find the UPGMA tree or the Neighbor-Joining tree for the set of organisms for which we want to build a tree. We can compute the tree length for this tree and use it as an upper bound under the reasonable assumption that the maximum parsimony tree will have a tree length less than or equal to this.

We now need to start identifying the possible trees. Here we use a branching method. We start with a tree consisting of three of the organisms. There are three ways of adding a fourth organism to this tree. Each of these three trees can then be extended in five different ways in order to add a fifth organism. Each of these fifteen trees can be extended in seven different ways if we wanted to add a fifth organism resulting in a total of 105 trees. You can see how the number of trees has started growing quite fast. Here is where the bound comes in. At each stage, we use the Fitch algorithm to compute the tree length and compare it against the bound. If the tree length is greater than the bound we don't extend it any further. So, suppose in the first step one of the three trees with four leaves has a tree length that exceeds the bound. In the next step, we only extend the other two resulting in ten rather than fifteen trees. If half of these fail the bound test we only extend five trees resulting in a total of thirty-five trees instead of 105 trees.

The branch and bound method does not affect the optimality of the final tree—we will still get the best tree. However, if our bound is not tight enough we might still end up evaluating a large number of trees. This method can significantly reduce the number of trees that need to be evaluated. However, the significant reduction is from a huge number so even after reducing the number of trees the maximum parsimony approach is computationally very expensive.

6.5.3 Maximum Likelihood

Instead of just counting the steps from the putative ancestor to the descendants we can also look at the likelihood of each tree. Computing the likelihood involves tracing the path from the ancestor to the descendants and multiplying the transition probabilities as we work through the tree. Let's again use the four-taxon example but this time look at a rooted tree with the four taxa as shown in Fig. 6.15. Notice that we have labeled the interior nodes of the tree as well as the leaf and root nodes. We will denote the base at node q as x_q and the transition probability from base q to r as $P_{x_q x_r}$. Then the likelihood of this tree for a particular column n in the multiple alignment is given by [8]:

$$L_n = P_{x_o} P_{x_o x_e} P_{x_e x_a} P_{x_e x_b} P_{x_o x_f} P_{x_f x_c} P_{x_f x_d}$$

In order to obtain the probabilities we need to use a model of some kind. Using the Poisson model we get

$$P_{ij} = \begin{cases} g_i + (1 - g_i)e^{-v} & i = j \\ g_j(1 - e^{-v}) & i \neq j \end{cases}$$

By picking appropriate values of g_i and v we can show that this is the same as the Jukes–Cantor distance.

Fig. 6.15 A candidate rooted
tree

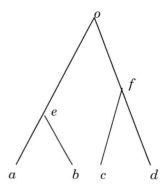

There is one problem with computing the likelihood as we have described. The bases at
the interior and root nodes are not available. This means that we will need to sum over all
possible values at these nodes. So the likelihood equation becomes:

$$L_n = \sum_{x_o} \sum_{x_e} \sum_{x_f} P_{x_o} P_{x_o x_e}(e) P_{x_e x_a}(a) P_{x_e x_b}(b) P_{x_o x_f}(f) P_{x_f x_c}(c) P_{x_f x_d}(d)$$

We compute this likelihood for each possible tree and pick the tree with the maximum
likelihood. Clearly, again we have the problem of computational cost.

6.6   Phylogenetic Networks

While we have concentrated on the hierarchical relationship between species, this approach
ignores the horizontal transfer of genetic material between species. This horizontal genetic
transfer (HGT) is an especially important feature of evolution in microbial communities and
has become an important source of threat to human health. The transfer of genetic material
is a possible pathway for the development of antibiotic-resistant bacteria. In bacteria, the
horizontal transfer of genetic material is made easier by the existence of extra-chromosomal
genetic material, namely plasmids. As bacteria are universal scavengers it is easy to see how
they could ingest foreign DNA. The incorporation of genetic material into the chromosome
can be achieved through the use of transposable elements. And the transfer of genetic material
among species can be facilitated by viruses.

In order to identify which parts of the DNA are horizontally transferred one needs to
look at ways of characterizing the genome which will allow us to differentiate between
"own" and "foreign". One way to do that is to use the idea of signatures. The process of
evolution over long periods of time has left its imprint on the DNA of individual organisms.
Each species has in some context a unique environment and thus one would expect that the
imprint of that environment would itself be unique. Where that imprint becomes pervasive
and discriminatory we call it a signature. By pervasive we mean that this imprint is evident

not just from the whole genome but in fragments of the genome and by discriminatory we mean that the imprint is different for genomes of different species. The size of the fragment which is sufficient to exhibit the signature makes the signature a weak signature or a strong signature. One of the best-known signatures is the GC content of a genome [4]. This is the ratio of Guanine and Cytosine nucleotides to the total number of nucleotides in a genome. It is not much of a signature, especially in eukaryotes where the GC content of the genome varies along the genome. In prokaryotes, the GC content does have some discriminatory capability. For example, bacteria in the Staphylococcus genus have lower GC content than bacteria in the Pseudomonas genus. A somewhat stronger signature is the dinucleotide frequency [5] where you construct a vector of the frequency of occurrence of different dinucleotides. These turn out to be different for different genomic groupings. A generalization of this is the idea of oligonucleotide frequencies [6]. Finally, a strong signature is the Average Mutual Information profile [7] which calculates the information contained in one base about another base k bases away. We can construct profiles of different lengths. The similarity between these profiles can be used to generate phylogenetic relationships [7].

6.7   Exercises

1. What is the definition of homology for gene sequences? Describe different types of homology.
2. List the main tree construction methods in phylogeny?
3. How does molecular phylogeny differ from classical phylogeny?
4. What are the two main weaknesses of the UPGMA algorithm?
5. How does the Neighbor Joining (NJ) algorithm differ from UPGMA in selecting pairs to join?
6. State and explain the molecular clock hypothesis.
7. Why is "computing the fraction of mismatches between two molecular sequences" an underestimate of "the actual evolutionary distance" between the two sequences?
8. Assume we are given the following multiple alignment for S1, S2, S3, and S4. Consider the tree topology shown on right. Write the likelihood for the highlighted column for two possible sequence of internal node assignments.

S1: A G C C T C A T C T
S2: C G C C T C A T A T
S3: A G T G G T A T A T
S4: A G T C G G C T A G

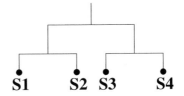

9. For column 2 and column 5 in the multiple alignment for sequences P, Q, R, S, and T, consider the tree shown next to it. Find the required mutations to generate the columns, i.e., their parsimony score. Determine and report the optimum internal node assignments in your calculation.

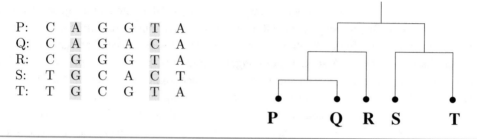

P: C   A   G   G   T   A
Q: C   A   G   A   C   A
R: C   G   G   G   T   A
S: T   G   C   A   C   T
T: T   G   C   G   T   A

P   Q   R   S   T

References

1. E. Zuckerkandl and L. Pauling. Molecular Disease, Evolution, and Genic Heterogeneity. In M. Kasha and B. Pullman, editors, *Horizons in Biochemistry*, pages 189–225. Academic press, 1962.
2. Richard E Dickerson. The structure of cytochromec and the rates of molecular evolution. *Journal of Molecular Evolution*, 1(1):26–45, 1971.
3. Walter M Fitch. Toward defining the course of evolution: minimum change for a specific tree topology. *Systematic Biology*, 20(4):406–416, 1971.
4. Giorgio Bernardi. The isochore organization of the human genome. *Annual review of genetics*, 23(1):637–659, 1989.
5. S. Karlin, A.M. Campbell, and J. Mrazek. Comparitive DNA Analysis across Diverse Genomes. *Annual Review of Genetics*, 32:185–225, December 1998.
6. S. Karlin and L.R. Cardon. Computational DNA Sequence Analysis. *Annual Review of Microbiology*, 48:619–654, 1994.
7. M. Bauer, S.M. Schuster, and K. Sayood. The average mutual information profile as a genomic signature. *BMC Bioinformatics*, 9, January 2008.
8. J.P. Huelsenbeck and K.A. Crandall. Phylogeny estimation and hypothesis testing using maximum likelihood. *Annual Review of Ecology and systematics* (1997):437–466, 1997.